AF841311

Offshore Safety Management

Offshore Safety Management

Implementing a SEMS Program

Ian Sutton

ELSEVIER

AMSTERDAM • BOSTON • HEIDELBERG • LONDON • NEW YORK • OXFORD
PARIS • SAN DIEGO • SAN FRANCISCO • SINGAPORE • SYDNEY • TOKYO

William Andrew is an imprint of Elsevier

William Andrew is an imprint of Elsevier
225 Wyman Street, Waltham, MA 02451, USA
The Boulevard, Langford Lane, Kidlington, Oxford OX5 1GB, UK

First edition 2012

Copyright © 2012 Ian Sutton. Published by Elsevier Inc. All rights reserved.

No part of this publication may be reproduced or transmitted in any form or by any means, electronic or mechanical, including photocopying, recording, or any information storage and retrieval system, without permission in writing from the publisher. Details on how to seek permission, further information about the Publisher's permissions policies and arrangements with organizations such as the Copyright Clearance Center and the Copyright Licensing Agency, can be found at our website: www.elsevier.com/permissions.

This book and the individual contributions contained in it are protected under copyright by the Publisher (other than as may be noted herein).

Notice
Knowledge and best practice in this field are constantly changing. As new research and experience broaden our understanding, changes in research methods, professional practices, or medical treatment may become necessary.

Practitioners and researchers must always rely on their own experience and knowledge in evaluating and using any information, methods, compounds, or experiments described herein. In using such information or methods they should be mindful of their own safety and the safety of others, including parties for whom they have a professional responsibility.

To the fullest extent of the law, neither the Publisher nor the authors, contributors, or editors, assume any liability for any injury and/or damage to persons or property as a matter of products liability, negligence or otherwise, or from any use operation of any methods, products, instructions, or ideas contained in the material herein.

Library of Congress Cataloging-in-Publication Data
A catalog record for this book is available from the Library of Congress

British Library Cataloguing in Publication Data
A catalogue record for this book is available from the British Library

ISBN: 978-1-4377-3524-6

For information on all Elsevier publications
visit our web site at elsevierdirect.com

Printed and bound in Great Britain

12 13 14 10 9 8 7 6 5 4 3 2 1

Working together to grow
libraries in developing countries

www.elsevier.com | www.bookaid.org | www.sabre.org

ELSEVIER BOOK AID International Sabre Foundation

Contents

Preface

Deepwater Horizon changed everything.

Before that April 2010 accident—in which 11 men died, record quantities of oil flowed into the ocean, and nearly $1 billion dollar's worth of investment plunged to the seabed—the safety record of the offshore oil and gas industry was good, and had been steadily improving. Which was why Deepwater Horizon was such a shock—it was so unexpected.

One of the most important consequences of the event was that the Minerals Management Service (MMS) that had had up to that time the overall responsibility for enforcing offshore safety went through a fundamental reorganization, and renamed itself the Bureau of Ocean Energy Management and Regulatory Enforcement (BOEMRE). And one of the first actions of the new agency was to issue the Safety and Environmental Management Systems (SEMS) rule.

But the impact of Deepwater Horizon goes beyond the United States—the accident caused oil companies all over the world to think through the effectiveness of their safety management programs. Moreover, events such as the Montara blowout in Australian waters in the year 2009 showed that these events are not confined to one place.

This book, *Offshore Safety Management*, has been written for those working in this new world—the world created by Deepwater Horizon. The book starts by describing the management of safety offshore. Topics covered include a brief history, the types of safety, risk management, and the role of regulations. The book then describes many of the major events and accidents that have occurred over the last 40 years or so, some of which occurred onshore, that led to the development of modern safety management systems and regulations.

The discussion then moves on to the Safety and Environmental Management Program (SEMP—the standard issued by the API in the early 1990s) and the newly published SEMS. These topics are front and center stage for any company operating in US waters. But they possess universal relevance. The principles and application of the elements of SEMS can be applied to any production platform or drilling rig anywhere in the world.

In Europe, Australia, and other locations around the world, companies in the offshore oil and gas business manage safety through the use of safety cases. This description is followed by a discussion of Formal Safety Assessments, which analyze technical topics such as fires, explosions, noise, and emergency escape.

This book should be read in conjunction with *Process Risk and Reliability Management*, published by Elsevier in 2010. That book provides much more detail regarding the implementation and management of the elements of SEMS such as operating procedures and mechanical integrity.

This book concludes with a discussion to do with likely developments in offshore safety management. The statement that "Deepwater Horizon changed everything" goes beyond a recognition that existing safety programs such as SEMS and safety cases need to be improved—it forces managers and professionals to consider new ways of thinking, including the need for leadership as distinct from management and an understanding of the long-tail or Black Swan events.

As always when I write, the most difficult challenge I face is knowing when to stop. When I first arrived in New York I was taught the phrase "Enough is enough already." There is always a good time to stop writing—and that time is now.

I must, of course, acknowledge the support of many colleagues working in the offshore safety management business, in particular the support provided by my employer AMEC Paragon in encouraging me to write articles, attend conferences, and participate in professional society committee meetings. And, as always, my deepest thanks to my wife Val for her patience and support—particularly at those times when I was writing instead of helping around the house.

Finally, in every publication that I have written so far I have concluded the preface with the words Edmund Spenser used in the introduction to his poem Faerie Queene, 'Goe little book: thy selfe present.' So it is with this book—it is now in your hands gentle reader. I hope you find it useful.

Other books on the topic, seminars and further information for the Process Industries can be found on the author's website: http://www.stb07.com/bookshop/book-offshore-safety-management.html

Warning—Disclaimer

This book provides information in regard to the subject matter covered. It is distributed with the understanding that the publisher, the author, and any other advisors are not engaged in rendering legal, accounting, or other professional services. If legal or expert assistance is required the services of a competent professional should be sought.

It is not the purpose of this book to reprint all the information that is available to the author and/or publisher, but to complement, amplify, and supplement other texts. You are urged to read all the available material, learn as much as possible about the topics covered, and to tailor the information provided here to your needs. For further information and other sources, see the references provided at www.stb07.com/citations.html.

Every effort has been made to make this book as complete and as accurate as possible. However, **there may be mistakes**, both typographical and in content. Therefore this text should be used only as a guide and not as an ultimate source of information and guidance on the topics covered. Furthermore, information on these topics may have changed since the release date of this book.

The examples provided, which are fictional, have been created specifically for this book series. This book does not present information about any specific plant, process, company, or individual. Nor does it describe the policy of any company.

The purpose of this book is to provide information in the subject areas covered. The author and publisher of this book shall have neither liability nor responsibility to any person or entity with respect to any loss or damage caused or alleged to be caused, directly or indirectly by the information and/or guidance provided in this book.

CHAPTER

Offshore safety management

1

CHAPTER CONTENTS

INTRODUCTION

The production of oil and gas offshore is risky. The danger of fires and explosions is always present and the toxic gas hydrogen sulfide (H_2S) is a frequent byproduct of the oil and gas that is produced. Drilling rigs and platforms are generally crowded with equipment and machinery, and—if there is an accident—there are few places to which to escape. Moreover, in locations such as the North Sea, weather conditions can be harsh and unforgiving.

Thus, from its very beginning, this industry has had to pay particular attention to the safety of the workers and the prevention of catastrophic events. Generally, these efforts have been successful, particularly with regard to occupational or "hard hat" safety. However, as discussed in Chapter 2, major accidents do occur quite frequently, and the costs of such accidents in terms of human life, environmental damage, and economic loss can be enormous.

This book has been written to help those who design and operate offshore facilities to build and improve on the industry's good safety record, and to take actions that will help prevent and control catastrophic events. Specifically, the book shows how to

Ian Sutton: Offshore Safety Management. DOI: 10.1016/B978-1-4377-3524-6.00001-0
Copyright © 2012 Ian Sutton. Published by Elsevier Inc. All rights reserved.

develop, implement, and operate a Safety Management Systems (SMS). The primary purpose of such a system is to identify the hazards that could lead to a major event, to assess the risk associated with them, and then to implement the appropriate safeguards and corrective actions in order to ensure that such events do not actually occur, or that their consequences are minimized.

At the time of writing, new safety management regulations are being put into place in the United States as a result of the Deepwater Horizon incident (Chapter 2). Because of their topicality, these new regulations, known as the Safety and Environmental Management Systems (SEMSs), are discussed in detail in Chapters 4 and 5. Although these regulations apply only to the United States of America, the principles embodied within them are general in nature and can be applied to offshore installations worldwide. In other words, developing and implementing a SEMS is good practice, wherever a facility may be located.

VALUE OF SAFETY MANAGEMENT SYSTEMS

Given that the development, implementation, and maintenance of Safety Management Systems are expensive and time consuming, it is important to establish that they are providing value, and that safety is improving.

The section on Industry Trends (see below) shows that, over the last 15 years or so, occupational safety trends have been very impressive. But SMSs are more concerned with the control of catastrophic events, such as those described in Chapter 2. Given that such events happen only rarely, it is difficult to develop trend lines in the same manner.

However, it is probably fair to say that most professionals in the offshore oil and gas industry accept that the management of catastrophic events has improved, but not nearly as dramatically as it has for occupational safety. The reason for this cautious response may be that an SMS program by itself is not enough—it is a necessary but not sufficient condition. To be effective, the SMS has to affect the culture of the organization, and the way people behave and act. The effectiveness of the program also depends on the leadership and decisiveness of the company management, as discussed in Chapter 8.

The conclusion that can be drawn is that, if the catastrophic events are to be controlled, an SMS provides a necessary foundation. However, it is just a foundation—it also has to be implemented and it requires forceful leadership.

STRUCTURE OF THIS BOOK

This book is organized into eight chapters as listed in Table 1.1.

INDUSTRY TRENDS

The offshore industry has made great strides in improving safety. Figure 1.1 shows the progress that has been made in the USA (mostly the Gulf of Mexico).

The chart in Figure 1.1 was published by the US Mineral Management Service (MMS)—the government agency responsible for offshore safety regulation and

Table 1.1 Book Structure

Chapter	Overview
1 — Offshore safety management	This chapter provides an overview of safety management in the offshore oil and gas industry. Trends within the industry are discussed, along with an assessment of the impact of the recent Deepwater Horizon incident. A description of the different types of safety—occupational, process, and technical—is provided, along with a description of Safety Management Systems. An overview of regulations and special safety issues to do with offshore operations is provided.
2 — Major events	The development of safety systems is largely driven by lessons learned from incidents—particularly catastrophic events. Chapter 2 provides an overview of some of the incidents that have led to the development of new offshore safety standards and techniques. In addition to describing important offshore incidents, three events from other industries are also described because of the impact that they had on all industrial safety management programs.
3 — SEMP	In response to the Piper Alpha catastrophe of 1988, the American Petroleum Institute (API) published Recommended Practice 75. It calls on offshore operators to prepare and implement a Safety and Environmental Program (SEMP). This chapter describes and discusses the scope and content of a typical SEMP.
4 — SEMS	In 2010, in response to the *Deepwater Horizon* disaster, SEMP was made a legal requirement. The new rule, which is described in Chapters 4 and 5, is known as the Safety and Environmental Management Systems (SEMS).
5 — Implementing SEMS	Many companies have a lot of work to do before they are in compliance with the new SEMS rule. Not only do they have to complete their SEMP, the rule adds many other requirements. This chapter describes how to implement a SEMS program, with a particular focus on those companies that have a lot of catch-up work to do.
6 — Safety cases	SEMS and SEMP are not the only types of Safety Management Systems. Offshore facilities in other parts of the world, for example, develop through the use of safety cases. This chapter describes the structure of a typical safety case. Related topics, such as levels of acceptable risk, are also discussed.
7 — Formal safety assessments	A very brief overview of some of the technical analyses used to build a Safety Management System—primarily in the design phase—are provided in this chapter. An overview of the topic of risk quantification is also provided.
8 — Developments in offshore safety	This final chapter provides a brief overview of how offshore safety management is changing, and what developments could plausibly occur over the next few years. Also discussed in this chapter are the topics of leadership and long-tail events. Some concluding thoughts to do with leadership and the role of regulations are provided.

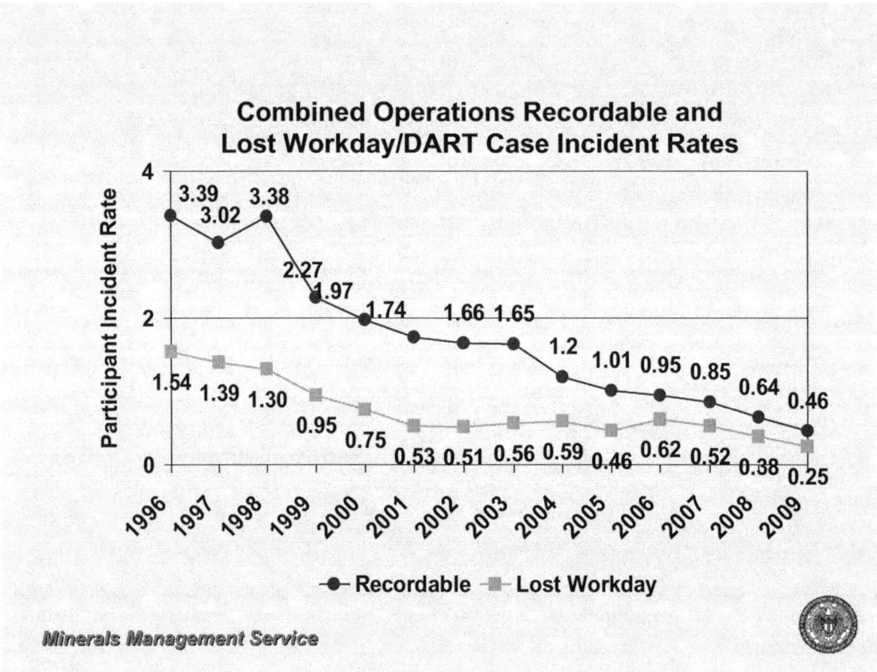

FIGURE 1.1

Safety Trends — USA Outer Continental Shelf.

enforcement. (Following the *Deepwater Horizon* incident, this agency was renamed the Bureau of Ocean Energy Management, Regulation, and Enforcement, or BOEMRE.) The chart provides data on the number of offshore safety incidents for the period 1996–2008. The trend is impressive: in just a 12-year period the recordable injury rate declined from 3.39 to 0.64: a drop of around 80%. The number of lost workdays dropped by a similar percentage. Moreover, the trend is quite smooth and steady, showing that the results are not a fluke or one-time event. And these advances were made as the industry has worked in ever-more challenging conditions—particularly as it has moved into very deep water operations.

The improvements shown in Figure 1.1 are not confined to the offshore industry. Similar trends are reported by onshore process facilities in the USA, as shown in Figure 1.2 (based on data from Pitblado, 2008).

The overall trend line in Figure 1.2, which is built on data from many large companies, demonstrates an order-of-magnitude improvement in occupational safety in the 12-year period covered. This is comparable to the trends shown in Figure 1.1.

Figure 1.3 shows a similar positive trend with respect to the industry's environmental record. The data, provided by the United States Coast Guard (USCG), include some land-based facilities, and also spills from inshore (State) waters. If the two bars for each of the years in Figure 1.3 are combined, it can be seen that the amount of oil spilled annually has declined from just under 6 million US gallons in the early 1980s to an almost negligible amount by the year 2005.

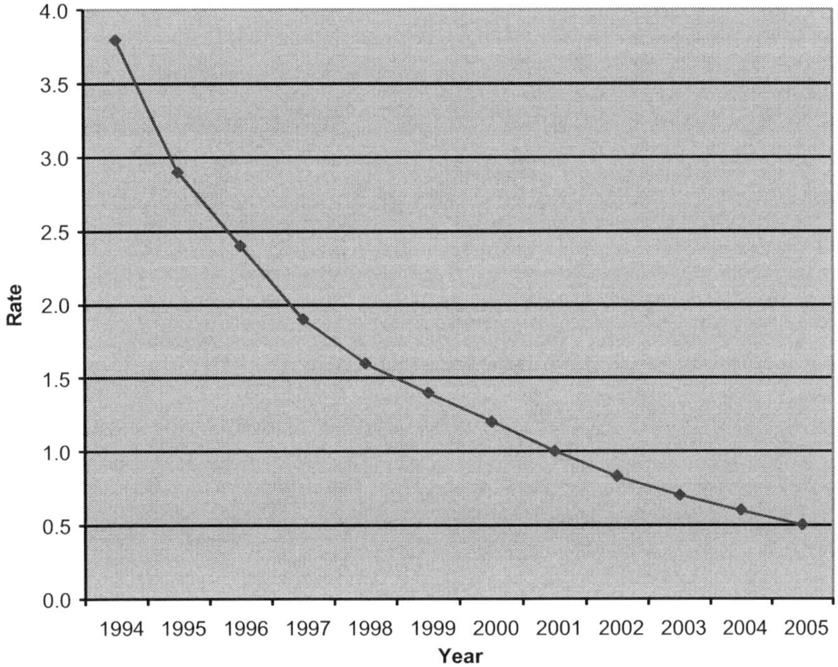

FIGURE 1.2

Safety Trends — USA Onshore Facilities.

So, up until the year 2010, the offshore oil and gas industry was able to demonstrate a good and steadily improving safety and environmental performance.

And then came Deepwater Horizon.

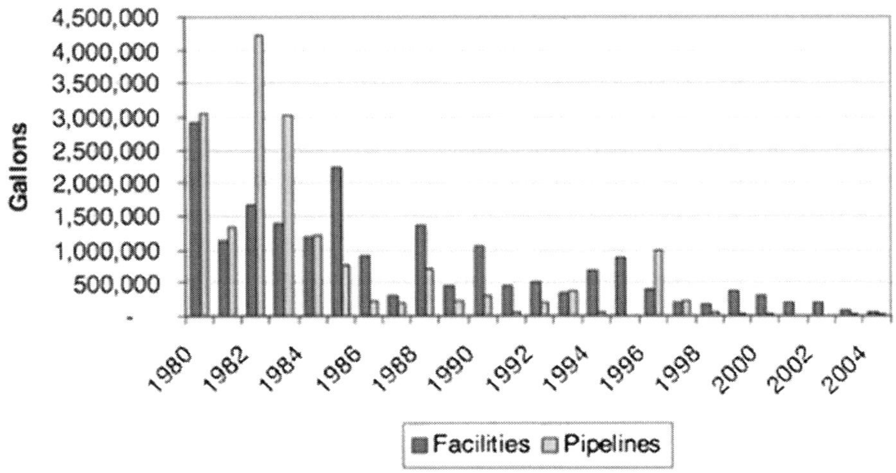

FIGURE 1.3

Environmental Trends — Offshore USA.

IMPACT OF DEEPWATER HORIZON

To say that the Deepwater Horizon catastrophe was a shock to the industry would be a gross understatement. Not only was the initial impact dramatic enough—the loss of life, the destruction of the rig and the massive spill—but the event showed that systems were not in place to respond promptly to a catastrophe of this magnitude. It took 5 months before the well could be sealed. And during those 5 months dramatic scenes were broadcast around the world showing oil pouring into the ocean 1 mile under the surface. Equally-compelling footage of events onshore, such as the death of wild birds and the closure of many small businesses, added to the impression of a situation out of control. In other words, the Deepwater Horizon event had implications that went way beyond the losses associated with just one drilling rig.

> *Deepwater energy exploration and production, particularly at the frontiers of experience, involve risks for which neither industry nor government has been adequately prepared, but for which they can and must be prepared in the future.*
>
> *National Commission (2011)*

The incident's 11 fatalities and the spill of approximately 150 million US gallons of oil will require the trend lines in Figures 1.1 and 1.3 to be rescaled. But catastrophic incidents such as this are often found to be not as much of an aberration as was thought at the time. They frequently reveal structural flaws in SMSs. Deepwater Horizon was no exception.

> *The oil industry has said the Deepwater Horizon rig catastrophe was a unique event, the result of an unprecedented series of missteps that are unlikely to be repeated. The recent history of offshore drilling suggests otherwise.*
>
> *Gold (2010)*

The above quotation from the *Wall Street Journal* cites a number of events and near-misses that occurred worldwide in a 12-month period prior to the *Deepwater Horizon* blowout. The frequency of such events suggests that the move to evermore-technically-challenging conditions (particularly drilling in deep water) have created problems that have not been fully overcome. Incidents that the article described included the following:

- In the Gulf of Mexico (GoM) there were 28 major well control spills or incidents in 2010. This number is up 4% from 2008, 56% from 2007 and over 60% from 2006.
- The number of serious incidents in the United Kingdom is up 31% from 2009 to 2010.
- The number of releases and well incidents in Norway is up 48% from 2009 (one of these—the Gullfaks C gas release—had the potential to be very serious).

Furthermore, the issues raised by Deepwater Horizon are not confined to the offshore oil and gas industry. At the time of writing, the consequences of the severe damage to the Fukushima nuclear power plants in Japan are still being ascertained. But it is clear that those consequences will be profound; substantial quantities of radioactive materials have been released, a significant fraction of Japan's power-generating capability is lost forever, and the cost of clean-up and remediation is going to be enormous. Indeed, the Fukushima accident may result in a massive slowdown in the construction of new nuclear power plants worldwide. Clearly, the offshore oil and gas industry is not the only one facing major challenges with respect to the management of safety and environmental performance.

SAFETY MANAGEMENT SYSTEMS

Many approaches to managing offshore safety are in use. Almost all of them require that an SMS is developed for each facility or group of facilities. Most of these systems, regardless of a facility's location or the legislative environment in which it operates, will be organized in a manner similar to that shown in Figure 1.4.

The six steps shown in Figure 1.4 are described briefly below, and are explained in greater detail in subsequent chapters of this book.

(1) Facility Description

The first step in the development of an SMS is to describe the facility for which the study and analysis is being conducted. Items to be considered include the following:

- The physical location of the facility;
- Its function (production, drilling or pipeline transportation);
- The organization that owns the facility, and which is responsible for its safe operation;
- The role of contractors and their relationship with the owner/operator;
- The connections (both physical and organizational) that the facility has with other facilities;
- The management of contractors and the way in which their safety programs are integrated with that of the owner/operator; and
- The regulatory regime in which the facility operates.

FIGURE 1.4

Structure of a Safety Management System (SMS).

(2) Technical Information

Technical information provides the foundation for most of the analyses and studies that comprise an SMS. Such information includes Piping and Instrument Diagrams, Layout Diagrams, and Drilling Plans. Guidance to do with the development and use of Technical Information is provided in Chapters 3 and 4.

(3) Risk Assessment

Once the facility description is complete and technical information is gathered, the next step in the development of an SMS is to conduct a risk assessment. Such an assessment consists of the following five steps:

(i) Identify the hazards;

(ii) Evaluate the consequences (safety, environmental, and economic) of those hazards should they materialize;

(iii) Estimate the predicted frequency of the hazards;

(iv) Determine the effectiveness of the system safeguards; and

(v) Assess the overall level of risk.

(4) Risk Acceptance

Management has to decide if the calculated level of risk is acceptable and, if it is not, what actions need to be taken to reduce it. Some SMSs—in particular many safety cases—are built around the concept of a numerical value for As Low as Reasonable Practicable (ALARP) value. If the risk lies above/below the predetermined ALARP value then corrective actions must be taken.

Other SMSs (including SEMP and SEMS) do not discuss acceptable risk *per se*, nor do they provide numerical values for acceptable risk. Instead they are largely driven by the acceptable levels of risk that are implicit in industry standards from bodies such as the American Petroleum Institute (API) and the American Society of Mechanical Engineers (ASME).

For example, the following text from API's *Recommended Practice for Analysis, Design, Installation, and Testing of Basic Surface Safety Systems for Offshore Production Platforms (RP 14C)* states:

> *The safety system should provide two levels of protection to prevent or minimize the effects of an equipment failure*

In this example, an expert committee has decided that two levels of protection provide an acceptable level of risk control. In effect, the determination as to what level of risk is acceptable has been transferred from the facility's engineers and managers to an expert committee that has then developed a consensus standard.

(5) Report

The next step in the development of an SMS is to report on the results of the work that has been conducted up to that point. The word "report" is used rather loosely here because some systems, such as the process safety management (PSM) regulation from the Occupational Safety & Health Administration (OSHA) or the SEMS rule, do not require that companies submit an actual report. Still, the management of those facilities needs internal reports to advise them on the progress that they are making.

Other management systems, in particular safety cases, frequently require that a detailed report is prepared, showing how safety was analyzed and how it is to be managed. This report is frequently submitted to the regulators.

(6) Audit

All management systems must include an audit process. In the words of one plant manager, "There is always news about safety, and some of that news is bad." Audits, reviews, and gap analyses are needed to ferret out the bad news. The audit results are used as the basis for the next SMS iteration.

The development and implementation of a SMS never ends. Risk can never be low enough; improvements can always be made. Therefore, once the six steps shown in Figure 1.4 have been implemented, management will start the whole process over again—usually at the risk analysis and planning steps—in order to achieve ever-higher levels of safety and economic performance.

For facilities that are still in the design stage, the actions taken, particularly during the Risk Assessment step, will change for each iteration. For example, early evaluations of risk will use a Hazard Identification (HAZID) technique; later on, as detailed engineering information becomes available, the more comprehensive Hazard and Operability Study (HAZOP) method will be used.

HISTORICAL BACKGROUND

Safety programs are developed in response to incidents—particularly catastrophic events. Some of the events that led to the development of SMSs offshore are described in Chapter 2. An overview of some of the more important regulatory and standards-setting developments (offshore and onshore) is provided here.

Offshore

The development of formal SMSs for offshore oil and gas facilities can be said to have started with the Piper Alpha catastrophe that occurred in 1988. Offshore platforms had safety programs before that time, of course, but Piper Alpha ushered in a new and much more thorough approach to system safety.

Following the accident, an investigation was conducted by a committee headed by the Scottish High Court judge, Lord Cullen. The committee's report was highly critical of the safety cases that had been in place prior to the accident.

In response to the Cullen report, the offshore industry took two different tracks, as shown in Figure 1.5. Companies operating in the North Sea (and, later on, other areas of the world such as Australia) continued with the safety case approach, as shown in the bottom track of Figure 1.5, but radically improved the thoroughness and quality of the documents, and put in place more stringent measures to ensure that the recommended measures were actually implemented.

In the USA (principally the GoM) the response to the Piper Alpha incident was equally vigorous, but followed a different path. Rather than following the safety case approach, it was recommended that companies develop a Safety and Environmental Management Program (SEMP) as described in API Recommended Practice 75. Like

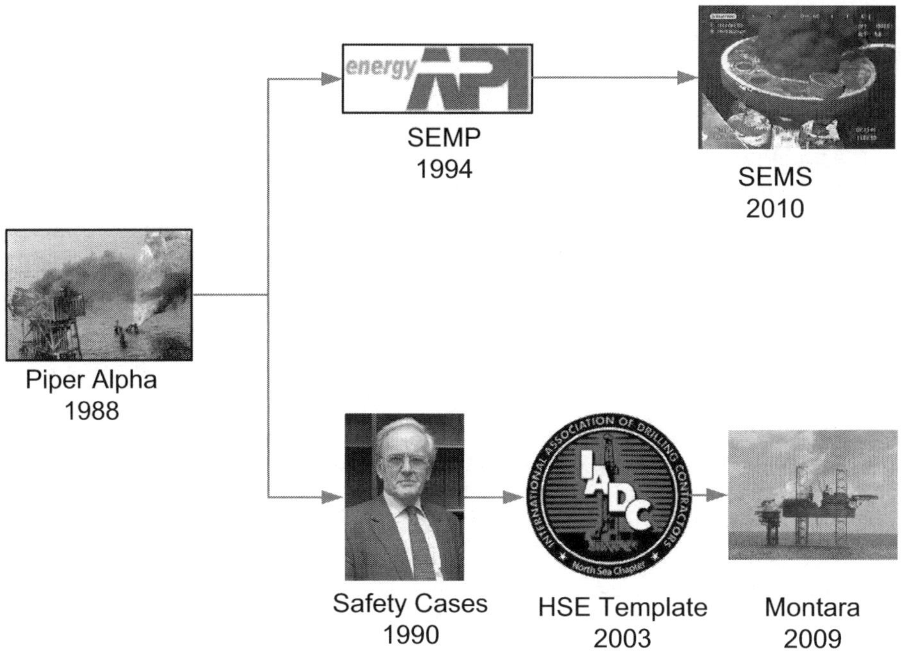

FIGURE 1.5

Impact of Piper Alpha.

a safety case, an SEMP is non-prescriptive. However, it makes extensive reference to industry standards (mostly from the API), and so is perceived as being considerably more prescriptive than the safety case approach. Nor does a SEMP require that a formal assessment of acceptable risk (ALARP) be determined.

Reasons for the non-use of safety cases in the USA include the following:

- The GoM has 5000–6000 platforms, many of them small and in shallow water. It is simply not economically feasible to write a safety case for each platform. Arnold (2010) uses the example of the response to a proposed requirement from the MMS to carry out a Failure Modes and Effects Analysis (FMEA) on each facility. Instead of doing this, a generic response based on FMEA and HAZOP analyses of 13 representative facilities was conducted. The use of a generic study was justified because the process equipment and instrumentation function in the same manner regardless of the specific design of the facility.
- The multiple small platforms are generally very similar to one another. Therefore, it makes more sense to develop universal standards (typically written and published by the API) than to conduct a formal analysis for each platform.
- The use of API standards and related documents has proven to be successful. The Deepwater Horizon incident was the first major release in the US waters since the Santa Barbara blowout of 1969, thus indicating that the SEMP-based systems have been effective. Furthermore, the recent Montara incident (discussed in Chapter 2) occurred in an area of the world that does use safety cases.

- The preparation of safety cases is time-consuming and involves a large amount of paper work. It is not clear if this administrative effort truly improves safety.
- When all platforms are designed and operated to the same standards, it is relatively easy to audit them. The auditor simply has to read the appropriate code or rule in order to determine compliance. Such is not the case with a safety case system, where each platform has its own unique program against which it has to be evaluated.

Onshore

The development and improvements in offshore safety discussed above and shown in Figure 1.1 were matched by corresponding changes and improvements in the onshore process industries in roughly the same timeframe.

Figure 1.6 provides a simple overview as to how regulations and standards in the onshore process industries developed.

In the 1980s, a series of serious incidents occurred in chemical plants throughout the world. With over 3500 fatalities, the Bhopal catastrophe of 1984 was the worst-ever industrial event in terms of loss of human life (although that event may well have been caused by sabotage). But there were many other high-profile incidents, including a sequence of explosions and fires in Pasadena, Texas, USA in the mid- to late-1980s.

Consequently, various nations enacted process safety-type legislation, such as the Seveso Initiative in Europe. In the USA, the Amendments to the Clean Air Act, which was

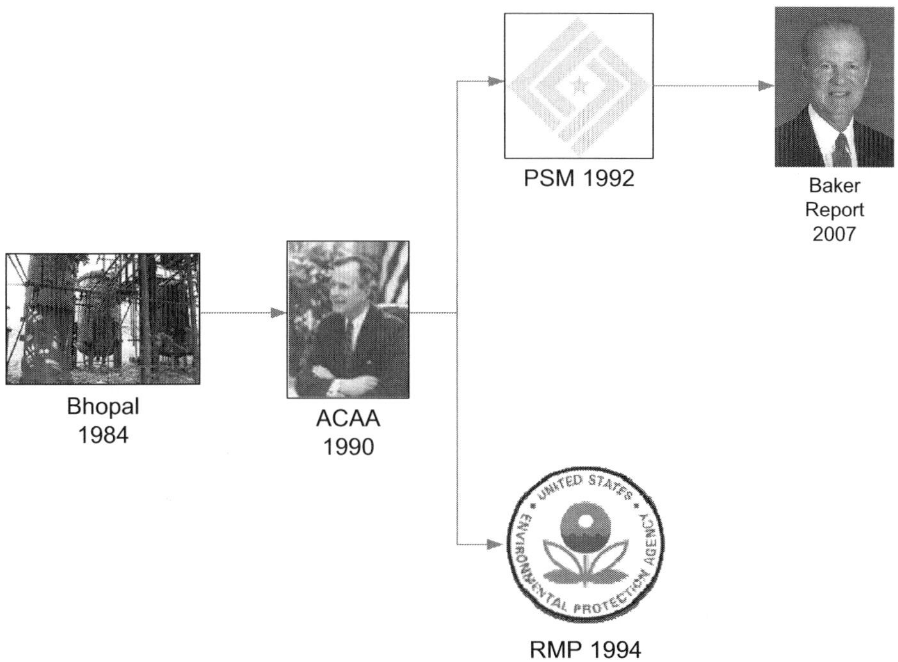

FIGURE 1.6

Onshore Safety Programs.

signed by President H.W. Bush just before he left office, required (amongst its many provisions) that the OSHA and the Environmental Protection Agency (EPA) put in place PSM and Risk Management Program (RMP) regulations. The Act was written such that the technical sections of the respective regulations are, however, very similar to one another, thus minimizing duplication of effort. (The legal and reporting sections of the two systems are very different from one another.)

OCCUPATIONAL, PROCESS, AND TECHNICAL SAFETY

Figure 1.1 at the beginning of this chapter showed that the number of offshore safety incidents in the USA has declined dramatically over the last 15 years. Although such a record is very creditable, it does beg the question as to how major events such as the Montara and Deepwater Horizon can happen. One answer to this question is to draw distinctions between Occupational Safety, Process Safety, and Technical Safety, and then to recognize that improvements in one area do not necessarily translate into improvements in the other areas.

Occupational safety

Occupational safety, sometimes referred to as "hard hat" or "trips and falls" safety, is concerned primarily with the behavior and performance of individuals, operating either alone or in small groups. It covers topics such as vessel entry, tripping hazards, vehicle movement, use of protective clothing, and lock out/tagout systems. It is what most members of the public think of when they hear the word "safety." Occupational safety incidents occur quite frequently, but their impact is generally limited to just a few people. Most of the improvements in safety shown in Figures 1.1 and 1.3 are in this area. This favorable trend has occurred because management and employees now give safety a much higher priority than they did, say, 20 years ago, and companies have invested heavily in a wide variety of programs such as the use of personal protective equipment (PPE), stop work authority, and first aid training. In particular, companies have emphasized the importance of behavior-based safety (Sutton, 2010a).

Process safety

Process safety focuses on process-related failures such as runaway chemical reactions, pipe ruptures, or vessel overflow. Such events may happen only rarely, but they can have catastrophic consequences. They are generally caused by failure to properly implement management systems such as operating procedures, mechanical integrity, training, pre-startup safety reviews, and management of change.

The datasets shown in Figures 1.1 and 1.2 are for occupational safety. It is much more difficult to find equivalent data for process safety events because catastrophic events occur only rarely. Anecdotal evidence suggests that there have been improvements in process safety in recent years, but not to the same extent as for occupational safety. The author of the paper in which Figure 1.2 is published states, "there is no clearly visible overall decline in major accident process safety events observed in either the USA or EU, although the data is noisy and some successes do exist—notably the UK Sector of the North Sea reduction in major leak events."

There are strong similarities between onshore PSM programs and the API's SEMP. This means that, because SEMP provides the basis for the SEMS rule, many of the practices and techniques that have been developed for onshore facilities can be used offshore with little modification.

OSHA's PSM standard

In the USA, one of the principal drivers for the implementation of PSM occurred in the onshore process industries in the early 1990s. Refineries and chemical plants handle large quantities of flammable and toxic materials, often at high pressure and temperature. There had been a series of serious process-related accidents in the 1980s, and it became apparent that new safety systems were needed. In the USA, the most important of these new systems was a regulation called the PSM of highly hazardous chemicals, 29 CFR 1910.119. It was promulgated in the year 1992 and is administered by the OSHA.

The management and technical elements of the OSHA PSM standard are shown in Table 1.2.

SEMP and SEMS

The offshore industry has two standards that are very similar to the OSHA PSM system described above. The first of these is the SEMP that is part of API RP 75. The second is the recently-introduced rule from BOEMRE—the SEMS. These standards are

Table 1.2 Elements of OSHA's PSM

(1)	Employee Participation
(2)	Process Safety Information
(3)	Process Hazards Analysis
(4)	Operating Procedures
(5)	Training
(6)	Contractors
(7)	Pre-startup Safety Review
(8)	Mechanical Integrity
(9)	Hot Work
(10)	Management of Change
(11)	Incident Investigation
(12)	Emergency Planning and Response
(13)	Compliance Audits
(14)	Trade Secrets

Table 1.3 Elements of SEMP/SEMS

(1)	Safety and Environmental Information
(2)	Hazards Analysis
(3)	Operating Procedures
(4)	Training
(5)	Pre-startup Review
(6)	Assurance of Quality and Mechanical Integrity of Equipment
(7)	Safe Work Practices
(8)	Management of Change
(9)	Investigation of Incidents
(10)	Emergency Response and Control
(11)	Audit of Safety and Environmental Management Program Elements
(12)	Records and Documentation

discussed in detail in subsequent chapters of this book. They both use a set of management elements very similar to those from OSHA, as can be seen by comparing Table 1.3 with Table 1.2. (The "General" part of SEMP has been excluded from Table 1.3.)

The Baker report

For the onshore industries the signature event that highlighted the differences between occupational and process safety was the explosion and fire at BP's Texas City refinery in 2005. For many years the refinery had a good safety record that was steadily improving. And then a massive explosion occurred. Many lives were lost, and there was extensive destruction of equipment and loss of production.

Following the event, an investigation was conducted by a committee chaired by Secretary James Baker (Baker, 2007). The following is a key quotation from that report.

> *Executive management tracked the trends in BP's personal safety metrics, and they understood that BP's performance in this regard was both better than industry averages and consistently improving. Based upon these trends, BP's executive management believed that the focus on metrics such as OSHA recordables... were largely successful. With respect to personal safety, that focus evidently was effective. BP's executive management, however, mistakenly believed that injury rates, such as days away from work case frequency and recordable injury frequency, were indicators of acceptable process safety performance... it was not until after the Texas City accident that management understood that those metrics do not correlate with the state of process safety.*

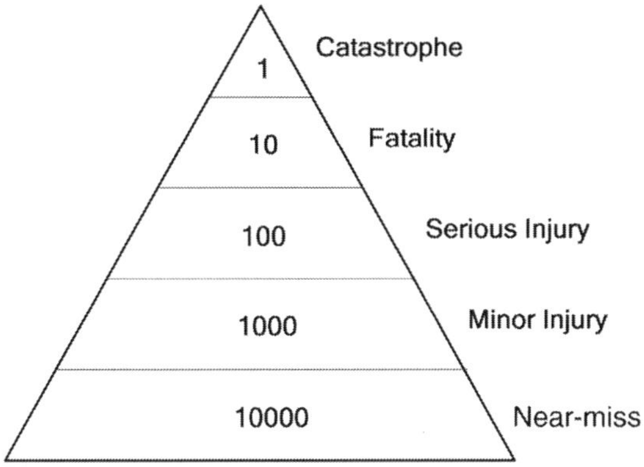

FIGURE 1.7

Incident Triangle — 1.

In other words, companies that have a good occupational safety record can still experience a catastrophic process safety related event. Improvements in personal safety do not necessarily reduce the chance of a major accident from occurring (although the reverse is less likely to be the case; a company that has a strong process safety program is likely to also have good occupational safety results). It is this difference between process and occupational safety that helps explain why process safety performance has not improved to the degree shown in Figures 1.1 and 1.2.

The incident triangle

One example as to how occupational and process safety need to be distinguished is to do with the much-quoted Incident Triangle, shown in Figure 1.7, which uses created information.

The basic idea behind the triangle is that serious events such as fatalities, large environmental spills, and serious financial losses occur only rarely. By contrast, near-misses and low-consequence events are much more common and can be seen as being precursors to the more serious events. If a relationship exists between the two types of events, then programs that reduce the number of near-misses and minor injuries will, it is argued, lead to a corresponding reduction in the number of catastrophes.

Figure 1.7 shows five levels of seriousness concerning worker safety (similar categories can be used for environmental and economic loss). Single order of magnitude steps are used. Hence it is estimated that, for every 10,000 near misses there will be 1000 minor injuries, 100 serious injuries, 10 fatalities, and 1 catastrophic event.

Various studies report on actual ratios. For example, Mannan et al. (2005) give the following ratios:

Fatalities	1
Serious injury	7
Minor injury	44
No injuries	300

The assumption underpinning the incident pyramid is that the causes for all types of events are the same. In fact, this assumption is only partially correct because the root causes of minor events are different from those that lead to process safety events. Therefore, improving "day-to-day" safety will not necessarily reduce the number of serious incidents. Minor events are typically caused by *occupational* problems such as trips and falls, lack of proper PPE, and improper use of machinery. Major events, however, are more often caused by *process* safety problems such as incorrect instrument settings, corrosion, or mixing of incompatible chemicals. Hence, a program that leads to improvements in occupational safety will not necessarily help reduce the frequency of process-related events. Indeed, improvements in the occupational safety record may induce a false sense of confidence regarding the potential for a major event. (It is probable, however, that a poor performance in occupational safety will correlate positively with a poor performance in process safety.)

The reason that good occupational safety performance does not necessarily correlate with the frequency of serious accidents is that occupational accidents tend to have simple causes. For example, if a worker traps his or her fingers in a piece of moving machinery, some likely recommendations that result from such an event are as follows:

- Ensure that that item of machinery, and all others like it, is properly guarded.
- Ensure that all affected personnel are properly trained in the use of that type of machine.
- Issue better PPE.

However, a thorough incident investigation into such an apparently simple accident could lead to the discovery of significant and subtle deficiencies in the overall management program that could, in turn, lead to ways of improving process safety.

Figure 1.7 can, therefore, be modified as shown in Figure 1.8. A disconnect is shown between the large number of occupational injuries and the much smaller number of catastrophic events that are process related.

Evidence as to whether an opposite, top-down effect may apply is hard to come by; that is, whether improvements in process safety lead to matching improvements in occupational safety. Many process safety professionals feel that such a trend does exist although it is difficult to prove.

Technical safety

The third category of safety analysis is Technical Safety. It focuses on topics such as the dispersion of toxic gases, the design of escape routes and fire, and the impact of explosions. The category has much in common with process safety, but they are not identical. The best time to conduct technical safety work is when a facility is being designed. For this reason, Technical Safety is sometimes referred to as Safety in Design.

There is a good deal of overlap between process and technical safety. However, process safety tends to focus on the role of people on facilities that are already in operation. Technical safety has more to do with technical analysis, and is less concerned with the behavior of people or the implementation of management systems.

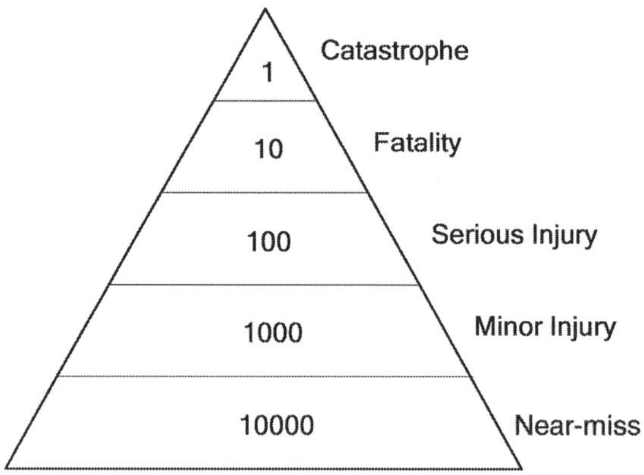

FIGURE 1.8

Incident Triangle – 2.

RISK MANAGEMENT

Risk management lies at the root of all process safety work. Risk is composed of the following four elements:

(1) A hazard;

(2) The consequences of that hazard should it occur (safety, environmental, and economic);

(3) The likelihood of occurrence of the hazard; and

(4) Safeguards that reduce consequences and/or likelihood.

The relationship between the first of the above three terms is shown in Eqn 1.1.

$$\text{Risk}_{\text{Hazard}} = \text{Consequence} \times \text{Predicted Frequency} \qquad (1.1)$$

The first and most important step in any RMP is to identify the hazards because, unless hazards are identified, consequence and likelihood reduction cannot be implemented. Not only is HAZID the most important part of any risk analysis, removal of hazards is almost always the best way of reducing risk, and it is the only way in which risk can be reduced to zero. Reductions in the consequence and frequency terms—the second and third elements of Eqn 1.1—can only reduce risk; they cannot eliminate risk entirely.

Perception of the consequence term

Equation 1.1 is linear; it gives equal value to changes in the consequence and frequency terms, implying a linear trade-off between the two. For example, according to Eqn 1.1, a hazard resulting in one fatality every 100 years has the same risk value as a hazard resulting in 10 fatalities every 1000 years. In both cases the fatality rate is 1 in 100 years, or 0.01 fatalities per year. But the two risks are not *perceived* to be the same. In general, people feel that high-consequence events that occur only rarely are less acceptable than more frequent, low-consequence accidents. Hence, the second of the two alternatives shown above is perceived as being worse than the first.

The same way of looking at risk can be seen in everyday life. In a typical large American city around 500 people die each year in road accidents. Although many efforts are made to reduce this fatality rate, the fact remains that this loss of life is perceived as a necessary component of modern life, hence there is little outrage on the part of the public. Yet, were an airplane carrying 500 people to crash at that same city's airport every year, there would be an outcry. Yet, the fatality rate is the same in each case, that is, 500 deaths per city per year. The difference between the two risks is a perception rooted in feelings and values.

To accommodate the difference in perception regarding risk, Eqn 1.1 can be modified so as to take the form of Eqn 1.2.

$$\text{Risk}_{\text{Hazard}} = \text{Consequence}^{n} \times \text{Likelihood} \qquad (1.2)$$

where $n > 1$.

Equation 1.2 shows that the contribution of the consequence term has been raised by the exponent n, where $n > 1$. In other words, high-consequence/low-frequency accidents are assigned a higher *perceived* risk value than low-consequence/high-frequency accidents.

Since the variable n represents subjective feelings it is impossible to assign it an objective value. However, if a value of say 1.5 is given to n then Eqn 1.2 for the two scenarios just discussed—the airplane crash and the highway fatalities—becomes Eqns 1.3 and 1.4, respectively.

$$\text{Risk}_{\text{airplane}} = 500^{1.5} \times 1 = 11\,180 \qquad (1.3)$$

$$\text{Risk}_{\text{auto}} = 1^{1.5} \times 500 = 500 \qquad (1.4)$$

The 500 airplane fatalities are perceived as being equivalent to over 11,000 automobile fatalities; that is, the apparent risk to do with the airplane crash is 17.3 times greater than for the multiple automobile fatalities.

Performance-based/prescriptive programs

RMPs can fall into one of two categories: prescriptive or performance-based. (In reality most programs are a mix of both.) Prescriptive programs are those whereby the success of prescriptive programs is measured, at least in the short term, by compliance with relatively detailed rules. One difficulty with the prescriptive approach is that technology changes very fast, particularly in deep water work, whereas the writing of rules and regulations is a slow and painstaking process. This means that prescriptive standards may not be sufficiently up to date to address current issues. Such a problem does not occur with non-prescriptive programs, such as safety cases. The management of the risk is the responsibility of the organization that creates the risk. If the organization has developed the technology that creates the risk, then that same organization can create the risk management systems that are needed to control the risk.

The use of prescriptive standards does, however, offer a number of advantages. First, given that the standards were developed by experts in the field, their use will ensure that high levels of safety will be achieved, even if the persons designing and running the platform are not themselves industry experts.

Second, the use of prescriptive standards increases efficiency and reduces design time. Rather than having to develop safety concepts and standards from scratch, the designers and operators of a platform can quickly and efficiently apply recognized rules.

Finally, a prescriptive system allows for facilities to be audited more quickly and more consistently. The quality of the audit does not depend as much on the training and knowledge of the auditor as it would in a non-prescriptive environment. Moreover, when all platforms are designed and operated to the same standards (mostly from the API) it is relatively easy to audit them. The auditor simply has to research the appropriate code or rule, and he or she can come to a quick conclusion. Such is not the case with a safety case system, where each platform has its own unique program against which it has to be evaluated.

Non-prescriptive management programs are always performance-based because the only measure of success is success. And success can only be achieved by not having incidents. But, from a theoretical point of view, such a goal is impossible to achieve. No matter how well-run a facility may be, accidents will occur; risk can never be zero. Accidents can always occur. For this reason an organization that follows a non-prescriptive RMP can never be "in compliance." Do whatever it takes *on your facility* not to have accidents. It is up to the managers, the technical experts, and the operations/maintenance personnel to determine how this should be done. (Of course, detailed rules do have to be followed when they apply; the safety case is not a justification or excuse for avoiding compliance.)

Acceptable risk

Given that risk is basically subjective, it is not possible to dispassionately define what level of risk is acceptable and what is not. What is acceptable to one person may be totally unacceptable to another. Regulatory agencies in particular will never place a numerical value on human life and suffering because any number that they develop would inevitably generate controversy. Yet, working targets have to be provided, otherwise the facility personnel do not know what they are aiming towards.

Fewtrell and Bartram (2001) provide some parameters for acceptable risk determination. They include the following:

- An arbitrary defined probability;
- The cost of reducing the risk is greater than the costs saved;
- The money would be better spent on reducing risks elsewhere;
- The public (and politicians) agree on the acceptable level of risk.

All of these parameters pose problems. For example, if the cost of reducing the risk is greater than the costs saved, then, rather than declaring this to be the acceptable level of risk, it may be more responsible simply not to carry out that particular activity. Similarly, saying that the money could be better spent elsewhere may simply indicate that a bigger budget is needed, not that a level of acceptable risk has been defined.

Formal Safety Assessments (FSAs) are built around the concept of "acceptable risk." For example, the decision as to whether a firewall should be installed between two sections of the platform will depend on the associated risk, that is, the nature of the hazard (spreading fire, consequence of the fire, and likelihood of its occurrence). The risk evaluation will then be repeated assuming the presence of a safeguard (in this case the firewall).

For this example, it may be determined that the risk moves from one fatality every 10 years to one every 1000 years, as illustrated in Figure 1.9.

In this simple example, the level of acceptable risk has been set at 0.01 fatalities per year, that is, one fatality every 100 years.

Many companies use a value of 0.001—one fatality every 1000 years. This is known as the Individual Risk per Annum (IRPA). A more complex term is Potential Loss of Life (PLL). It combines IRPA with the number of people present to come up with a forecast as to how often a fatality would occur on that facility.

Given the legitimate difficulties in defining acceptable risk, and given the legal exposure that could ensue should a company do so, many companies and organizations simply choose to punt, that is, they discuss the concept purely qualitatively and avoid assigning numerical values. One large oil company, for example, has an internal ALARP guide that is over 30 pages long. Yet, at no point in the document is any numerical guidance provided for acceptable risk. Essentially, the document was one long circular argument on the following lines:

- Is the risk low enough?
- What is the threshold for low risk?
- The threshold for low risk depends on a perception that the risk is low enough.

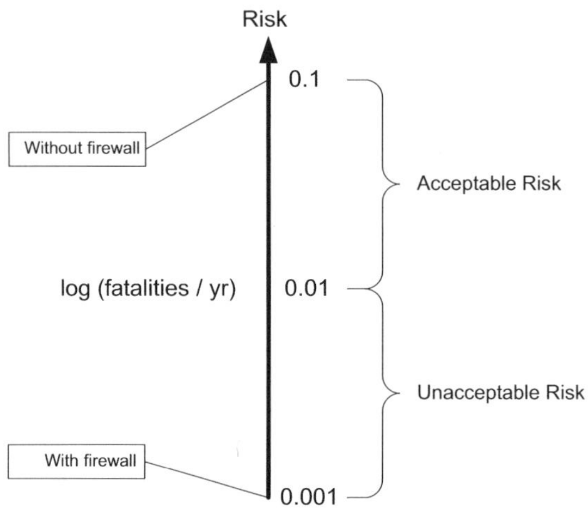

FIGURE 1.9

Acceptable Risk.

For example, the designers of an offshore platform have to decide on the length of time that the temporary refuge will survive in the event of an explosion followed by a continuous fire. If the design team settles on a time period of say 30 minutes, then they have implicitly made a decision as to the level of acceptable risk.

Uncertainty

Related to the above difficulties are concerns to do with the accuracy of any risk predictions. Analysts can carry out calculations that result in the predictions—yet, the quality and accuracy of such calculations are only very approximate, at best.

Cost–benefit analysis

Discussions regarding acceptable risk usually boil down to matching the value of human life and suffering with the cost of implementing additional safety measures. Such discussions are fundamentally futile: there is no way in which an objective value can be assigned to human loss.

Nevertheless, any risk analysis must make such a comparison, even if such a comparison is implicit.

Risk reduction

Regardless of how risk levels are determined, the United Kingdom Health and Safety Executive suggests that the hierarchy for risk reduction shown below should be followed:

- Eliminate and minimize the size of hazards using the principles of inherent safety design;
- Reduce the likelihood of the event;
- Detect the occurrence of the event;
- Control the impact of the event (scale, intensity, and duration);
- Mitigate the consequences; and
- Implement EER (Emergency, Escape, and Rescue)

Risk matrices

Risk is commonly analyzed and managed through the use of a system of three risk matrices. They are as follows:

- Consequence matrix;
- Frequency matrix; and
- Risk matrix.

Consequence matrix

A representative consequence matrix is shown in Table 1.4. The matrix has four levels of consequence covering worker safety, public safety, environment, and economic loss. There are no rules as to how many levels should be selected, nor does

Table 1.4 Consequence Categories

	Worker Safety	Public Safety	Environment	Economic (Annual)
Low, 1	Reportable or equivalent	None	Limited impact that is readily corrected	\$10,000–\$100,000
Moderate, 2	Hospitalization or lost-time injury	Minor medical Attention	Report to Agencies and take corrective action	\$100,000–\$1 million
Severe, 3	Single disabling injury	Hospitalization or serious injury. Some local reporting	Irreversible damage to low quality land, or clean-up of environmentally sensitive areas required	\$1 million–\$10 million
Very Severe, 4	Fatality or multiple serious injuries	Fatality or multiple serious injuries. Massive negative publicity	Months of clean-up work needed in environmentally-sensitive areas	≥\$10 million

any major regulatory body insist on a particular size of matrix. However, many companies choose four levels; three levels do not provide sufficient flexibility and differentiation, but five levels imply a level of accuracy that is probably not justified. The steps in Table 1.4, from "Low" to "Very Severe," are roughly in order of magnitude, that is, each increased level is about 10 times more serious than the one before it.

Frequency matrix

Once the consequences associated with an incident have been identified, the next step is to estimate the frequency with which the incident may occur. A representative frequency matrix is shown in Table 1.5. As with the consequence matrix, four value

Table 1.5 Frequency Matrix

	Frequency	Comments
Low, 1	<1 in 1000 years	Essentially impossible: "Once in a blue moon" or "meteor falling out of the sky"
Moderate, 2	1 in 100 years to 1 in 1000 years	Conceivable—has never happened in the facility being analyzed, but has probably occurred in a similar facility somewhere else
High, 3	1 in 10 years to 1 in 100 years	Might happen in a career
Very High, 4	>1 in 10 years	It is likely that the event has occurred at the site if the facility is more than a few years old

levels are provided. The use of just three levels is probably too coarse, but five levels or more implies a degree of accuracy that probably could not be justified (precision is not the same as accuracy).

As with the consequence matrix, the steps in Table 1.5 are roughly an order of magnitude greater than the one before it.

Risk matrix

Having determined consequence and frequency values for a particular hazard, the overall risk is determined using a third matrix such as that shown in Table 1.6, which shows four levels of risk.

The risk values will usually line up diagonally, with all the values in any one diagonal being the same.

The meaning of the four letters in Table 1.6 is as follows.

A — Very High

This level of risk requires prompt action; money is no object, and doing nothing is not an option. An "A" risk is urgent. On an operating facility, management must implement immediate temporary controls (ITCs) while longer-term solutions are being investigated. If effective ITCs cannot be found, then the operation must be stopped. During the design phases of a project, immediate corrective action must be taken in response to an "A" finding, regardless of the impact on the schedule and budget.

B — High

Risk must be reduced, but there is time to conduct more detailed analysis and investigations. Remediation is expected within, for example, 90 days. If the resolution is expected to take longer than this, then an ITC must be put in place.

C — Moderate

The risk is significant. However, cost considerations can be factored into the final action taken, as can normal scheduling constraints such as the availability of spare parts or the timing of facility turnarounds. Resolution of the finding must occur within approximately 18 months. An ITC may or may not be required.

D — Low

Requires action but is of low importance. In spite of their low risk ranking, "D" level risks must be resolved and recommendations implemented according to a schedule;

Table 1.6 Risk Ranking Matrix

	Consequence			
	Low, 1	**Moderate, 2**	**Severe, 3**	**Very Severe, 4**
Low, 1	D	D	C	C
Moderate, 2	D	C	C	B
High, 3	C	C	B	A
Very High, 4	C	B	A	A

they cannot be ignored. (Some companies do allow very low risk-ranked findings to be ignored on the grounds that they are within the bounds of acceptable risk.)

As low as reasonably practicable risk—ALARP

To provide a measure for allowable risk, companies use the concept of "As Low as Reasonably Practical" risk. This term is usually abbreviated to ALARP (sometimes facetiously referred to "As Low as Regulators Permit"). Other terms that have the same general meaning are "As Low as Reasonably Achievable" (ALARA) and "So Far as is Reasonably Practicable" (SFAIRP).

Setting ALARP

Due to the subjective nature of risk, no external agency, whether it be a regulatory body, a professional society or the author of a book such as this, can provide an objective value for ALARP. What risk is acceptable is very much in the eye of the beholder.

Nonetheless, individuals and organizations are constantly gauging the level of risk that they face in their personal and work lives, and then acting on their assessment of that risk. For example, at a personal level, an individual has to make a judgment as to whether it is safe to cross a busy road. In industrial facilities, managers make risk-based decisions regarding issues such as whether to shut down an equipment item for maintenance or to keep it running for another week; or a manager may have to decide if a proposed change should be scrutinized through the use of a full hazard and operability (HAZOP) analysis. Engineering standards, and other professional documents, can provide guidance. Indeed, for well-understood risks in standard situations the application of industry codes and standards is generally sufficient to reduce risks to ALARP. But, regardless of what guidance is provided, at the end of the day, the manager has a risk-based decision to make. That decision implies that some estimate of "acceptable risk" has been made.

When quantified values for ALARP are not available, the following can help establish levels of acceptability:

- Use of the best available technology capable of being installed, operated, and maintained in the work environment by the people prepared to work in that environment.
- Use of the best operating and maintenance management systems available.
- Maintenance of equipment and management systems to a high standard.

On a project, the ALARP process will become more detailed as the project design progresses. During the early stages of the project, the focus will be on identifying the major hazards, assessing their associated risks and determining which of the three ALARP techniques discussed above is to be used. In later stages of the design, barriers will be selected and it will be shown that they reduced risk to an acceptable level.

Difficulties with ALARP

A fundamental difficulty with the concept of ALARP is that the term is inherently circular and self-referential. For example, the phrase "best available technology" used in the list above can be defined as that level of technology which reduces risk to an acceptable level—in other words to the ALARP level. Terms such as "best operations" and "high standard" are equally ambiguous.

It may be for reasons such as these that the UK HSE in the year 2006 decided to minimize its emphasis on ALARP requirements from the safety case regime for offshore facilities. Some companies have also elected to move away from ALARP toward a continuous risk reduction model (Broadribb, 2008).

Reverse ALARP

Occasionally, industry managers have attempted to show that the risk mitigation measures that are in place are too stringent, and that they could be reduced (with the consequent savings), without crossing into the area of unacceptable risk.

Needless to say, the regulators have not looked favorably on such propositions. It is hard to defend a situation in which an operator deliberately makes safety worse.

REGULATIONS

> Note: This section contains thoughts and discussion regarding legal issues. The material was written by an engineer. Therefore, it goes without saying that a qualified attorney should be consulted regarding specific problems and issues.

Because the offshore industry has such a potential for catastrophic events it is heavily regulated. Although regulations are often seen as being burdensome and onerous they do help ensure that safety and environmental standards are maintained, and that all companies are held to the same level of performance. The existence of regulations is particularly important in those situations where a company may be tempted not to bother investing in safety. For example, a company that is attempting to wring the last drops of oil out a depleted well may have less enthusiasm (and capital) for implementing a full SMS than if it were drilling a new and large prospect. Regulations help prevent corner-cutting in situations such as these.

Types of safety regulation

Regulations can be categorized in the following ways:

- Whether they are prescriptive or goal-based;
- The manner in which it enforced;
- The reporting requirements of the agency.

The approach selected will depend on factors such as the overall regulatory culture and the level of technology being applied.

Prescriptive/Goal-based

Safety regulations can be either prescriptive or non-prescriptive (goal-based). Prescriptive regulations are what most people think of when safety is discussed, and are often associated with traditional occupational safety standards such as the height of hand-rails or the use of PPE. To pick an example at random, OSHA has the following rule about ladders:

> *All rungs shall have a minimum diameter of 3/4 inch for metal ladders, except as covered in paragraph (b)(7)(i) of this section and a minimum diameter of 1 1/8 inches for wood ladders.*

A statement such as that is quite clear and uncompromising. It is also easy to follow. A person installing a ladder at an industrial site need not think about the basic principles of ladder design or use. Nor does that person have to carry out any type of risk analysis. He or she simply has to follow the rule regarding ladder design.

However, the prescriptive approach to rule-making has three fundamental problems when it comes to the prevention of major accidents:

(1) The causes of major incidents are typically complex, and involve a series of events—many of which are of low probability. Realistically, there is no way that a regulator can write a rule or standard to cover such situations.

(2) The operation of modern offshore oil and gas platforms involves the use of sophisticated technical and management systems and technology. It is unlikely that a regulator, no matter how well versed he or she may be in the offshore oil and gas business, will possess sufficient knowledge to be able to write and enforce regulations that keep up with all of the latest technical developments.

(3) There is no means of designing or operating to a calculated level of risk. Even if it could be showed that the use of a different type of ladder in this example is of very low risk, the OSHA rule has to be followed.

Therefore, the management of risk offshore tends to use a more non-prescriptive or goal-based set of standards. Such standards do not provide specific instructions as to what needs to be done. The basic idea behind them is that the companies that operate sophisticated facilities are the ones who know the process and equipment the best, so they are the best qualified to determine what needs to be done to achieve safe operations. The regulator then judges their performance on results achieved. Basically, a non-prescriptive rule says, "Do whatever it takes on your facility not to have accidents." It is up to the managers, the technical experts, and the operations/maintenance personnel to determine how this should be done.

Because non-prescriptive standards are performance-based, it is never really possible to be "in compliance" because any company, no matter how good its safety management may be, can have an incident.

In practice, most non-prescriptive regulations also incorporate many prescriptive standards. For example, the SEMS rule that is described in Chapter 4 is basically non-prescriptive, but it does make reference to standards such as API RP 14C that is quite prescriptive in nature. For example, paragraph 4.2.1.1.4 of the SEMS rule states,

> *Secondary protection from overpressure in a pressure component should be provided by a PSV. Secondary protection for atmospheric components should be provided by a second vent.*

Although more nuanced than the OSHA rule about ladders, the above sentences are just as prescriptive.

Enforcement mechanism

In addition to the prescriptive/non-prescriptive distinctions, rules and regulations also differ in the manner in which they are enforced. Broadly speaking, enforcement can be either external or internal (self-regulatory). External enforcement generally comes from agencies such as BOEMRE or OSHA. Internal enforcement is applied by the facility itself.

There is overlap between the two approaches. For example, a company that uses self-regulation always faces the possibility of an external audit from a government agency. Also, its program will be scrutinized very carefully indeed by outsiders if there is a serious accident. Some companies use internal auditors that are supplied by corporate headquarters. They act in the same way as external auditors.

Reporting requirements

A final distinction between the types of standard is the requirements for providing a report to an external agency or the public. Some regulations, such as the RMP rule from the EPA, require that the covered facility provide a detailed report describing the SMS (safety cases are another example). Other standards, such as OSHA's PSM, require no report at all. The system can be audited, but a facility does not have to submit a report on its program to the agency.

Examples of types of regulation

Figure 1.10 shows how some of the regulatory systems discussed in this book fit into the structures just described. Although useful for illustrative purposes, it is important to recognize that there are many similarities between the different management systems, and hence a good deal of overlap between them. A sketch such as that shown in Figure 1.10 does not reflect such nuances.

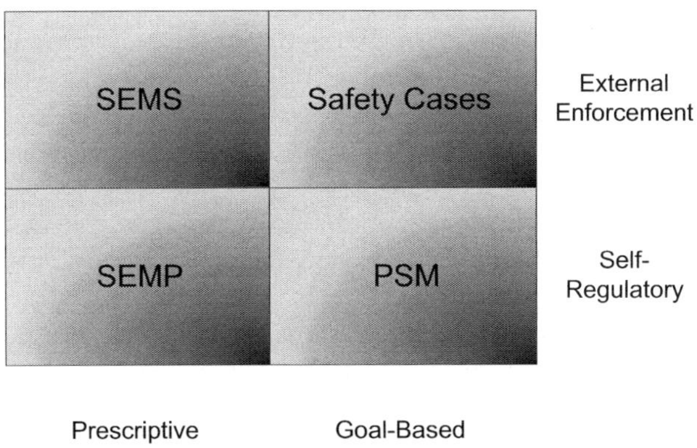

FIGURE 1.10

Types of Regulation (with Examples).

The *top left* square in Figure 1.10 covers those standards that are, broadly speaking, prescriptive in nature and enforced by outside parties. SEMS—a rule that is discussed in detail in Chapter 4—fits this category. Although built on a goal-setting base (the SEMP standard) it refers to many prescriptive standards and is enforced by an outside agency (the BSEE, part of BOEMRE).

The *bottom left* square in Figure 1.10 is for those safety programs that are prescriptive in nature, but where it is up to facility management to make sure that the standards are applied. SEMP fits this category. From a technical point of view it is very similar to SEMS. However, it relies entirely on companies enforcing their own systems.

The *top right* square is for goal-based programs that are enforced by outside parties. Safety cases are an example of this type of program. In principle, they are self-enforced because they represent the case that management makes *to itself* that the facility is safe. In practice, safety cases have been incorporated into law and are enforced by external agencies such as the UK HSE or the Australian National Offshore Petroleum Safety Authority (NOPSA). The development and use of safety cases is discussed in Chapter 6.

The *bottom right* square covers goal-based programs are not subject to external regulation. Onshore PSM programs fall into this category. The PSM standard from OSHA does not provide much detailed guidance; nor does it refer to other standards in the way that SEMP does. Moreover, the PSM standard is self-regulatory in that a company's program does not have to be evaluated or approved by OSHA or any other agency. OSHA's only involvement is to conduct periodic audits, and to participate in the investigation if a serious incident takes place.

The regulator's dilemma

From a regulator's point of view it is fairly simple to measure compliance against a prescriptive standard. Either a facility is following that standard, or it is not—there is not much judgment involved. For example, with regard to the API RP 14C standard quoted above, either a system has two levels of protection, or it does not. In the case of the OSHA ladder standard, either the rungs of a ladder have a minimum diameter of $^3/_4$ inch, or they do not. There is no need for judgment.

With regard to non-prescriptive, goal-based rules, however, the regulator faces something of a dilemma. In these situations facility owners/operators develop their own safety programs that are designed to meet their specific circumstances. If a regulator approves those programs then he or she has implicitly stated that the program is satisfactory and that the assumed level of risk is acceptable. If, later on, a deficiency is found with the program (either during an audit or an incident investigation) then the regulator must take some responsibility (and the owners/operators can deny total responsibility).

The regulators over safety cases (Chapter 6) face the same dilemma because, if there were to be an accident, the company involved can claim that some of the responsibility for the event lies with the regulator. To get around this quandry, the UK HSE states that,

> *... "acceptance" requires satisfaction with the duty holder's approach to identifying and meeting health and safety needs. HSE "accepts" the validity of the described*

approach as being capable, if implemented as described, of achieving the necessary degree of risk control, but HSE does not confirm the outcomes of that approach.

Even participating in the development of a SMS be an issue. The Australian authority, NOPSA, had initially attended some of the operators' HAZID and HAZOP studies. However, they concluded that doing so compromise their independence.

... attendance by NOPSA OHS Inspectors at format project hazard identification and risk assessment meetings, such as HAZID/HAZOP meetings and the possible impact such attendance may have on stakeholders' perception of NOPSA's independence and its ability to provide assurance that risks are properly controlled in Australia's offshore safety regime.

Moreover, it is unlikely that the regulator will have a sufficient number of qualified people to fully evaluate the SMS for each and every offshore facility. Indeed, that is why companies prepare their own safety system—only they fully understand the technology and systems on the platform or rig.

Responsible parties

The regulators are not the only ones who face a dilemma regarding responsibility for results. Typically, there are multiple parties involved in the ownership and operation of an offshore facility, and, with regard to safety, it is not always clear as to who is responsible for what.

Offshore oil and gas properties are almost always owned by the national government. Through an agency such as BOEMRE the government then leases the property to an operator. In the case of the *Deepwater Horizon* incident, for example, BP leased the Macondo property from the US government. BP then became the operator. Typically, the operator arranges financing, is responsible for most of the costs of exploration and production, and collects the profits (or a royalty on the profits) from subsequent production. It is the operator who has ultimate responsibility for the safety and environmental performance of the facility, and also for making sure that all pertinent rules and regulations are followed. Sometimes the operator will delegate control of operations to a Duty Holder, who will have full operational responsibility for the work that is being done—including the facility's SMS.

Supporting the owner/operator is a multitude of contractors, carrying out a wide variety of tasks—some of which are critical to the facility's safety. In the case of *Deepwater Horizon*, for example, the drill rig itself was owned and operated by the company Transocean; other major contractors, such as Halliburton, were involved in making key decisions. It is probably not an exaggeration to say that, of all the challenges that offshore managers face with regard to safety, it is the relationship with contractors that poses the greatest problems.

RULE-MAKING PROCESS (UNITED STATES OF AMERICA)

In the USA, an agency wishing to introduce a new rule such as SEMS must publish that proposed rule for public comment. At the end of that time the agency holds a public hearing during which members of the public ask questions and make comments. Once the hearing is concluded, the agency provides feedback on all the comments that it has received (either verbal or written) and then publishes the final rule.

With regard to SEMS, this process was followed at first. The MMS published a draft rule containing just four of the technical elements of SEMP. They allowed time for written comments, and then held a public hearing in New Orleans in 2009. However, following the Deepwater Horizon catastrophe, the MMS/BOEMRE used its emergency powers to publish the final rule, which was much more comprehensive than the draft, without any additional public comments or additional hearings.

In an interview published by the *Houston Chronicle* on February 12, 2011, BOEMRE Director Bromwich stated that "his agency has no plans to make additional emergency rules. Rather, any changes will now go through the typical rule-making process that takes up to two years..." It is not anticipated that this statement will be applied retrospectively to the existing final rule.

REGULATORY AGENCIES (UNITED STATES OF AMERICA)

A brief overview of the US regulatory environment is provided here. (This material does not provide legal guidance, it merely provides technical readers with an overview of the regulatory environment.)

The regulatory picture for offshore operations in the US federal waters is difficult to follow because so many agencies are involved. These include the following in a rough order of importance:

(1) The Bureau of Ocean Energy Management, Regulation, and Enforcement (BOEMRE), and its subsidiary the Bureau of Safety and Environmental Enforcement (BSEE);

(2) The Coast Guard;

(3) The EPA;

(4) Department of Transportation (DOT);

(5) National Oceanic and Atmospheric Administration (NOAA);

(6) US Army Corps of Engineers.

The number of agencies, and the complexity of their relationships, makes their coordination challenging. (The situation onshore for refineries and chemical plants is much simpler.) For example, when a floating drilling rig is moving to its location, it is a marine vessel under the control of a captain who has to comply with Coast Guard standards. However, once the rig is on station it is under the control of an offshore installation manager (OIM) and under the jurisdiction of the BOEMRE.

Each authority was created by different laws that delegated different functions to different agencies at different times. BOEMRE, USCG, EPA, and DOT all have their own areas of jurisdiction, which sometimes overlap. For example, BOEMRE regulates production safety systems and related fire fighting, but USCG regulates personnel health and safety, lifesaving equipment, and fire fighting. Fire fighting jurisdiction is further complicated by a distinction between fixed and floating facilities. The EPA regulates air and wastewater discharges and waste disposal, but the USCG regulates marine pollution and garbage. The DOT regulates interstate pipelines and hazardous waste labeling, shipping, and handling.

When the authority of two or more agencies overlap they generally resolve the lines of responsibility by writing a Memorandum of Understanding (MOU). The negotiations leading to the creation of an MOU can be quite contentious.

Outer continental shelf

The first section of the extension of a continent's land mass under the ocean is referred to as the continental shelf. It is generally shallowly sloping and quite shallow (up to 200 m deep). It is followed, as shown in Figure 1.11, by the continental slope, the continental rise, and the deepwater ocean.

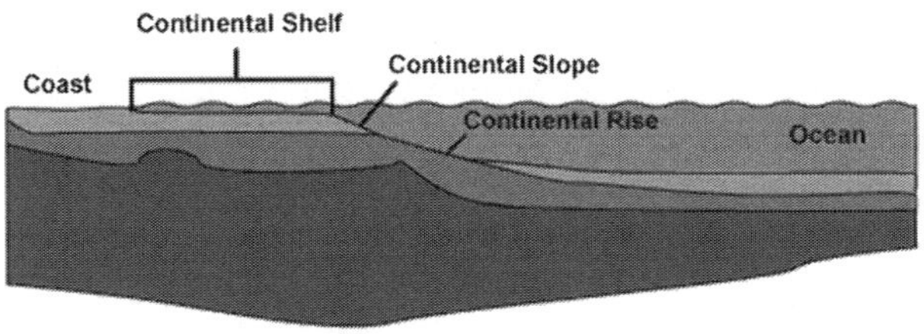

FIGURE 1.11

Continental Shelf.

Office of Naval Research

The Outer Continental Shelf (OCS) consists of the submerged lands, subsoil, and seabed, lying between the seaward extent of the States' jurisdiction (see Section 'State jurisdiction') and the seaward extent of Federal jurisdiction. The outward limit of federal jurisdiction is determined by international law. Determination of the distance depends on various factors, but a value of 200 nautical miles (370.4 km) is the base line.

State jurisdiction

In the United States, individual states have jurisdiction over the first few miles of the OCS that lies closest to them. The distances are as follows:

- Texas and the Gulf Coast of Florida: 9 nautical miles (16.7 km);
- Louisiana: 3 imperial miles (4.8 km);
- All others: 3 nautical miles (5.6 km).

Minerals Management Service

The MMS is the precursor organization to BOEMRE—the agency that currently manages offshore safety regulations in federal waters. The National Commission report (p. 77) to

President Obama that was written in the wake of the Deepwater Horizon disaster provides a useful overview of the creation of the MMS in the year 1982 also of and the tensions that existed from its very inception between its two functions: the collection of (very substantial) revenue for the US government, and ensuring that offshore operations were conducted safely. The Commission's report indicates that the move toward risk-based analysis that had occurred in most other nations was opposed both by industry organizations, and governmental bodies in the USA, in the years leading up to the Deepwater Horizon catastrophe.

BOEMRE

Following the Deepwater Horizon catastrophe the MMS was reorganized and renamed the Bureau of Ocean Energy Management, Regulation, and Enforcement, or BOEMRE. The agency is second only to the Internal Revenue Service as a source of funds for the US government. Given that the same agency is responsible for enforcing offshore safety, there is a clear potential for a conflict of interest. Therefore, one of the drivers regarding the structure of the new agency was that the revenue collection function was clearly separated from safety enforcement.

An independent report had already criticized the MMS with respect to highly improper behavior by some of its employees (in non-safety-related activities). The *Deepwater Horizon* added to the perception that the rules to do with safety were neither strict enough nor being enforced with sufficient rigor.

In September 2011, two sub-bureaus within the new agency were formed. The first of the new bureaus, the Bureau of Ocean Energy Management (BOEM) is responsible for managing development of the nation's offshore resources in an environmentally- and economically-responsible way. The second bureau, the Bureau of Safety and Environmental Enforcement (BSEE), is the one of most interest to safety professionals because it is the one that enforces safety and environmental regulations. Its functions include all field operations including "Permitting and Research, Inspections, Offshore Regulatory Programs, Oil Spill Response, and Training and Environmental Compliance functions." (In this book the terms "MMS," "BOEMRE," and "BSSE" are used to describe the agency at the time of the activity cited.)

Associated with the BSEE will be a permanent advisory body—the Offshore Energy Safety Advisory Committee (Safety Committee)—through which the nation's leading scientific, engineering, and technical experts will provide input on improving offshore drilling safety, well containment, and spill response.

Other regulatory changes include the following:

- Creation of a first-ever Chief Environmental Officer;
- Development of a new Environmental Compliance and enforcement function in BSEE;
- More prominent Oil Spill Response Plan review and enforcement in BSEE.

The Safety Committee will also advise the Director and the Secretary on a variety of issues related to offshore energy safety, including drilling and workplace safety, well intervention and containment, and oil spill response. The committee will facilitate

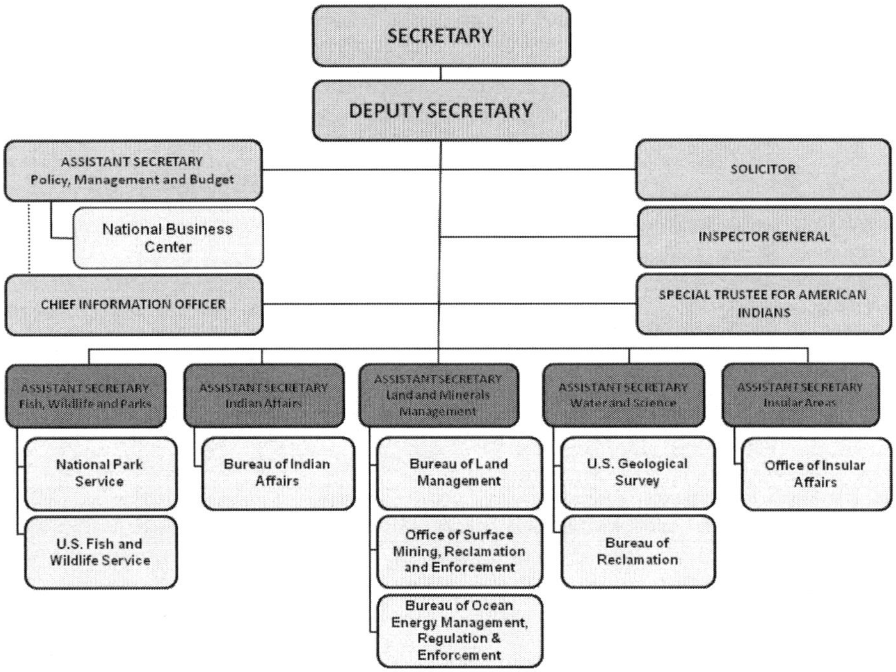

FIGURE 1.12

Organization of the Department of Interior.

collaborative research and development, training and execution in these and other areas relating to offshore energy safety.

BOEMRE is a part of the Department of Interior (DOI), as shown in Figure 1.12, which is taken from the Department's web site. It shows that BOEMRE is under the Assistant Secretary for Land and Minerals Management. Key leadership personnel are shown in Figure 1.13. (Figures 1.12 and 1.13 were current in the second quarter of 2011.)

BOEMRE is responsible for "overseeing the safe and environmentally responsible development of energy and mineral resources on the OCS." The agency is organized into three regions: Alaska, GoM, and Pacific.

One difficulty that has arisen with respect to the new name is how to pronounce the abbreviation "BOEMRE." In conversation the terms "BOE," "BOEM," and "BOEMRE" are all used. Probably the most widely used abbreviation is "BOE."

Strictly speaking, the abbreviation should be preceded by the word "the" because it is "the" Bureau. However, the agency itself uses the stand-alone term "BOEMRE" in the SEMS rule, so that is the convention that is adopted in this book.

Bureau of Safety and Environmental Enforcement

Of the three bureaus formed under BOEMRE, the one that is of most interest to those responsible for offshore safety and the development of SEMS programs is the BSEE. This office, which is generally pronounced "Bessie," handles the safety and environmental

Secretary — Ken Salazar

Deputy Secretary — David Hayes

Assistant Secretary, Land and Minerals Management — Wilma Lewis

Director Bureau of Ocean Energy Management, Regulation and Enforcement — Michael Bromwich

FIGURE 1.13

Organization of the staffing at the Department of Interior.

enforcement functions of the BOEMRE including, but not limited to, the authority to inspect, investigate, summon witnesses and produce evidence, levy penalties, cancel or suspend activities, and oversee safety, response, and removal preparedness. More detail as to the organization of this bureau was provided on January 19, 2011. In summary:

> *This office will be responsible for ensuring comprehensive oversight, safety, and environmental protection in all offshore energy activities. It will handle the safety and environmental enforcement functions of the Minerals Management Service (MMS) including, but not limited to, the authority to inspect, investigate, summon witnesses and produce evidence, levy penalties, cancel or suspend activities, and oversee safety, response, and removal preparedness. The bureau will have the authority to inspect, investigate, summon witnesses, produce evidence, levy penalties, cancel or suspend activities, and oversee safety, response, and removal preparedness. It expected that the bureau will have oversight over the development and implementation of the new Safety, Environmental and Management System (SEMS) rule.*

At the time of its formation in January 2011 the new bureau had approximately 300 employees (out of a total of 1700 employees from the disbanded MMS).

The formation of this Bureau was announced in January 2011. A fact sheet regarding its organization was published by BOEMRE in the same month.

30 CFR Part 250

BOEMRE rules are collected in the Federal Register under 30 CFR (Code of Federal Regulations) Part 250. The title of this code is "Oil and Gas and Sulfur Operations in the Outer Continental Shelf." It covers exploration, development, and production. Part (b)(2) of Section 250.101 summarizes its overall intent:

(b) Conform to sound conservation practice to preserve, protect, and develop mineral resources of the OCS to:

[...](2) Balance orderly energy resource development with protection of the human, marine, and coastal environments;

Some key sections of this Part regarding technical and process safety are:

- All platform production facilities must be protected with a safety system designed, analyzed, installed, tested, and maintained in operating condition in accordance with *API RP 14C — Safety Analysis for Production Platforms.*
- Fire fighting systems for both open and totally enclosed platforms shall conform to *API RP 14 G — Fire Prevention and Control on Fixed Open-type Offshore Production Platforms*.
- Hazards analyses should be carried out in accordance with *API RP 14 J — Design and Hazards Analysis for Offshore Production Facilities*.

The practical effect of requirements such as these is that some of the API Recommended Practices are now no longer just *recommended*, they have been incorporated into law. In other words, some of the "voluntary" SEMP standards have been legal requirements for many years.

Subparts

30 CFRP Part 250 is divided into the following subparts:

A. General

B. Plans and Information

C. Pollution Prevention and Control

D. Oil and Gas Drilling Operations

E. Oil and Gas Well-Completion Operations

F. Oil and Gas Well-Workover Operations

G. Reserved

H. Oil and Gas Production Safety Systems

I. Platforms and Structures

J. Pipelines and Pipeline Rights-of-Way

K. Oil and Gas Production Requirements

L. Oil and Gas Production Measurement, Surface Commingling and Security

M. Unitization

N. OCS Civil Penalties

O. Well Control and Production Safety Training

P. Sulfur Operations

Q. Decommissioning Activities

R. Reserved

S. Safety and Environmental Management Systems (SEMS).

It is the last of these—Subpart S—that is the focus of Chapters 4 and 5 of this book. Subpart O, training, is an important part of a SEMS program.

Each of the subparts has many sections. For example, under Subpart O (Training) is *§250.1501 What is the goal of my training program?*

Notices to lessees and operators

BOEMRE issues Notices to Lessees and Operators (NTLs) as a means of providing guidance. The NTLs are included in the Subpart information. The agency states the following about NTLs:

> *[they] clarify, supplement, or provide more detail about certain requirements. NTLs may also outline what you must provide as required information in your various submissions to BOEMRE.*

A bone of contention between BOEMRE and industry, particularly following the *Deepwater Horizon* event, is that the agency has been using NTLs in cases when full rule making may be called for. For example, the first version of the SEMS rule, which was published in draft form in 2006, contained just four of the elements of SEMP. A public hearing was held in 2009 in New Orleans. Yet, no further hearings were held before the final issue of the rule in October 2010, even though the number of elements increased from 4 to 12. Although the change included a discussion of the comments that had been made on the earlier version, the dramatic increase in scope may have justified a new hearing.

Inspections

One of the biggest challenges that BOEMRE faces is finding and training a sufficient number of qualified inspectors who are able to competently evaluate an SEMS program. The inspectors from the MMS traditionally worked with prescriptive standards. For example, as already noted, API RP 14C calls for two independent means of protection. Therefore, with respect to control of high pressure in a vessel, the inspector could verify that there was (a) a high-pressure interlock and (b) a pressure safety relief valve. Thus, the requirements of the standard are met, and the inspector is not required to exercise judgment.

SEMS is a different kettle of fish. Inspectors will be asked to evaluate systems such as management of change or operating procedures that are hard to define prescriptively. In the case of operating procedures, for example, an inspector using the prescriptive approach can determine if the procedures exist. But he or she will need considerable industry experience before being able to determine if those procedures are too long or too short.

To help address issues such as the above, BOEMRE plans on opening a National Offshore Training Center for its own people (Bromwich, 2011). Informal communications suggest that this center will be located in the state of Louisiana, USA.

Coast guard

The USCG is responsible for regulating the "safety of life and property on OCS facilities, vessels, and other units engaged in OCS activities." Its responsibility covers vessels such as Mobile Offshore Drilling Units, Drillships, and Floating Production Storage and Offloading Vessels (FPSOs) that work in the oil and gas industries. The USCG's activities cover the following:

- Fire fighting, lifesaving, personal health, and safety;
- Spill response oversight;
- MARPOL (control of trash and garbage).

One of the responsibilities of the USCG is to ensure that any collapsed, or partially collapsed, platforms are removed so as not to present a hazard to shipping. This is a particular concern following a hurricane when some of the older, inshore platforms may be knocked over by the winds and high seas. Such platforms are sometimes referred to as "Downers and Leaners" (see also p. 44).

Classification societies

Many of the requirements to do with marine standards are managed by classification societies. These are nongovernmental organizations that set technical rules, confirm that designs and calculations meet these rules, survey ships and structures during the process of construction and commissioning, and periodically survey vessels once they are in service. Their scope covers offshore structures.

A classification society will only state whether a vessel conforms to the pertinent codes and hence is either "in" or "out of" class. For liability reasons, the society will not state that a vessel is fit or unfit to sail. It simply reports on compliance with the required codes.

EPA

The EPA has responsibility for issues such as:

- Spill prevention;
- Wastewater discharges;
- Air emissions; and
- Solid and hazardous waste.

Operators are required to report to the EPA on any spills to the ocean.

DOT

The DOT has responsibility for subsea transmission pipelines, but not for pipelines directly connected to platforms. They also have some responsibility for hazardous waste management.

NOAA

The NOAA is concerned with the Federal Coast Management program and management of the Essential Fish Habitat.

Army Corp of Engineers

The Corp of Engineers has responsibility for shipping fairways and anchorage areas.

The Jones Act

The Jones Act (46 U.S.C. § 688(a)) allows injured sailors receive proper damages from their employers for the negligence of the ship owner, the captain, or fellow members of the crew. The act, which is part of the Merchant Marine Act of 1920 (updated in 2006), is also used to control the use of foreign vessels in US waters. For this reason, in the immediate aftermath of the Deepwater Horizon event, there were problems bringing in non-American vessels to help with the clean-up work.

SAMS (California)

The State of California has a Safety Assessment Management System (SAMS) for the control and management of offshore facilities. It falls under the jurisdiction of the California State Lands, and its enforcement relies to a large extent on a cooperative relationship between the State and offshore operators.

PFEER REGULATION (UNITED KINGDOM)

In the United Kingdom, many of the offshore regulatory requirements are handled through the Prevention of Fire and Explosion, and Emergency Response (PFEER) Regulations. They require that duty holders provide adequate arrangements for escape, evacuation, rescue, and recovery of persons involved in an incident.

INTERNATIONAL (SOLAS)

The International Convention for the Safety of Life at Sea (SOLAS) is an international maritime safety treaty. The first version of the treaty was passed as a result of the *Titanic* disaster in the year 1912 in which the largest passenger ship in the world hit an iceberg and sank.

SOLAS requires flag states to ensure that their ships comply with minimum safety standards in construction, equipment, and operation. Many of the SOLAS requirements have been adopted into the laws of various nations. It has the following 12 chapters:

Chapter I — General provisions

Chapter II-1 — Construction

Chapter II-2 — Fire protection, fire detection, and fire extinction

Chapter III — Life-saving appliances and arrangements

Chapter IV — Radiocommunications

Chapter V — Safety of navigation

Chapter VI — Carriage of cargoes

Chapter VII — Carriage of dangerous goods

Chapter VIII — Nuclear ships

Chapter IX — Management for the safe operation of ships

Chapter X — Safety measures for high-speed craft

Chapter XI-1 — Special measures to enhance maritime safety

Chapter XI-2 — Special measures to enhance maritime security

Chapter XII — Additional safety measures for bulk carriers

SPECIAL SAFETY ISSUES OFFSHORE

Earlier in this chapter it was noted that offshore safety management and the implementation of an SEMS program have much in common with other process industries, particularly onshore facilities such as refineries, pipelines, and chemical plants. However, each industry does have its own unique safety issues and problems. For example, a chemical plant will often either manufacture or use some highly toxic compounds that have to be handled with special equipment or materials of construction. Refineries have many processing steps that often operate at higher temperatures and pressures. And the pipeline business works in an environment where most of its operations are in public locations and close to where people live and work, so any type of leak is likely to quickly have a significant public impact.

The offshore business does not have to address most of the above problems. For example, offshore platforms use chemicals such as methanol and monoethylene glycol for hydrate removal, and although these chemicals are toxic and flammable, they are much less hazardous than many of the chemicals found in a typical chemical plant or refinery. Moreover, the quantities used offshore are quite small compared with a typical onshore facility. Generally, the chemicals are supplied in tote tanks that are offloaded from a supply boat and stored on a dedicated section of the deck.

In addition, operating conditions on offshore platforms tend to be less severe than those in chemical plants and refineries. Temperatures and pressures are usually quite low (downstream of high pressure chokes), and the process steps consist mainly of the physical separation of oil, water, and gas. Even if there is a chemical processing step, such as the removal of H_2S from the gas stream, the process is likely to be fairly simple—certainly when compared with the processes used in onshore chemical plants.

Finally, offshore platforms, unlike onshore pipelines, are pretty much out of the public eye. Unless a platform or rig has a very serious spill, offshore events are not likely to directly impact members of the public.

Nevertheless, the offshore business also has its own special safety issues. First and foremost, an offshore platform or rig is *offshore*. Hence all the hazards and dangers associated with marine work – dangers that stretch back over millennia – must be considered.

Specific issues to do with the safety of oil and gas offshore facilities include the following:

- Lack of escape routes;
- Persons on board;
- Cyclones/Hurricanes;
- Downers and leaners;
- Blowouts;
- H_2S;
- Dropped objects;
- Helicopters;
- Ship collisions.

These topics are discussed briefly below, and in greater detail in Chapter 7.

Lack of escape routes

On an onshore facility such as a refinery or chemical plant, workers at the site can, in the event of a major release or fire, escape from the scene (once they have secured the equipment for which they are responsible, and assuming that they are not part of the Emergency Response Team). They can literally run away. On one refinery an operator was drenched in light hydrocarbons that poured from a leak in a failed filter housing. He was surrounded by fired heaters, and he recognized that a fire was imminent and there was nothing he could do about it. So he simply ran from the scene. A major fire did erupt, and much of the equipment was seriously damaged, but no one was injured. Had he remained in the vicinity of the leak he would probably have died in the resulting fire (which was large and intense), and the resulting damage would have been no less.

Offshore, however, there is nowhere to run or to hide. Platforms are typically very congested so there is no "outside." If someone wishes to "run away" his or her only option is to enter the sea, preferably by a Totally Enclosed Motor Propelled Survival Craft (TEMPSC) or life raft. Going into the water directly may be feasible in warm locations such as the GoM or Angola, but jumping into the North Sea or the ocean off the coast of Labrador, particularly in winter, is likely to be a death sentence. Even in warm waters, hitting the surface of the sea can cause serious injuries, and there is no guarantee that a person will not hit part of the steel structure before landing in the water.

Persons on board

At onshore facilities, the people who are not at work go home. Therefore, in the event of a serious accident the number of affected people is limited to those who are on duty. For example, the explosion at the chemical plant in Flixborough, England, in the year 1974 was very bad: 28 men died. But the accident occurred on a weekend; had it occurred during the middle of a working day the death toll would have been much higher because the blast destroyed many of the office buildings at the plant site. (Actually, the number of deaths would have been zero had the operators, like the refinery operator described above, realized that they had lost control of the situation, and they needed to get away. As it was, they stayed in the control room for more than 20 minutes following the initial release; they paid for that decision with their lives.)

On an offshore platform, however, the persons who are not on duty are still present; and they may be killed or injured in the event of a catastrophe. Many of the deaths on the Piper Alpha platform (Chapter 2), for example, were of off-duty crew who were sleeping, and who could not escape from the living quarters. By contrast, 25 years later, when the explosion and fire occurred on the Deepwater Horizon platform, 11 men died instantly, but the other 135 persons on board survived the blast and subsequent fire. This is an indication that the safety measures that have been designed into rigs and platforms in the years following Piper have had a positive effect on safety improvement.

Hurricanes/cyclones

Hurricanes (also known as cyclones or typhoons depending on the geographic location) develop over warm bodies of water; they create high winds, thunderstorms, and high sea levels. Many offshore oil and gas facilities, particularly in the GoM, are vulnerable to hurricanes. Tropical storms and tropical depressions have the same causes as cyclones but generally do not generate such high winds, but they can create considerable rainfall.

Hurricanes in the GoM are given the names of persons. Four of the more important in terms of their impact on offshore facilities in recent years were Ivan (2004), Katrina (2005), Rita (2005), and Ike (2008). Some of the facts associated with these storms include the following:

(1) Sustained wind speeds were 175 mph (Katrina) and 180 mph (Rita). These are record values.

(2) Ivan generated large subsea mud slides. One platform is now buried under 50 ft of mud.

(3) Of the 4000 platforms in the GoM around 3050 were in the path of at least one of the storms. Of these, 113 platforms were lost and 52 were seriously damaged.

(4) Twenty-two of the downed platforms were less than 10 years old.

Downers and leaners

If a hurricane is severe enough to knock a platform over so that the entire structure lies on the floor of the ocean, then that platform is a "downer." If the platform partially collapses, so that some structure remains above the sea surface, then it is a called a "leaner."

These damaged platforms have to be removed because they pose a hazard to shipping. However, their removal is time-consuming and expensive. One operator had a cluster of about a dozen downers and leaners, all in the same general area and about 150 ft of water. The cost of removing the damaged structures was over $US1 billion. The wells associated with the platforms were producing very little oil and gas so it was decided to decommission them. Hence, there was not even any revenue for all the money that was spent.

Even when the water is shallow, it is not safe to send divers into a "downers and leaners" situation because the unstable structure may collapse while the diver is in the water. Moreover, if the wellhead has not been properly decommissioned, a gas blowout could occur at any time—once more posing a great risk to divers in the vicinity. This means that Remotely Operated Vehicles (ROVs) have to be used. Explosives are rarely used because of the potential impact on turtles and other marine creatures. Chemicals, including diesel fuel, that were on board the facility at the time of the platform collapse, pose an environmental hazard and can be a safety hazard for divers in the area.

It is often difficult to get construction information about older platforms. Either the information is missing or there has been a lot of "midnight engineering," that is, undocumented changes. In some cases the only records were on the platform that went down.

The "Reef in Place" program allows some platforms to be used as potential reefs. However, they must be at least 90 ft below the water line, and regulatory permission is required. Some companies are relocating their damaged platforms to other reefs in place locations.

Blowouts

As the recent *Deepwater Horizon* and *Montara* events have demonstrated so dramatically, blowouts can be both dangerous and environmentally destructive. They are also very expensive, in terms of both destroyed equipment and wasted production. Therefore, any SMS for a drilling rig must pay particular attention to the avoidance of blowouts.

A blowout occurs when the pressure of the hydrocarbons in a formation exceeds the pressure of the column of mud in the annulus of the drill string. Oil and gas rise very quickly up the string (with the gas expanding as it goes). When they reach the deck of the drill rig, they can ignite and/or create a major spill.

To prevent blowouts from occurring, the density of the drilling mud must be sufficiently high and the Blowout Preventer (BOP) must be functioning properly.

Hydrogen sulfide

H_2S is a highly toxic, colorless, flammable gas that is often found in crude oil—sometimes in the form of mercaptans, which break down to form H_2S. Not only is the gas very dangerous for human beings, it can also cause serious corrosion problems in piping and instruments.

Dropped objects

Dropped objects (usually from deck cranes) are a major hazard on offshore platforms. If they fall on the deck they can hurt workers and/or seriously damage equipment (with the potential for a catastrophic event). If the dropped object is heavy and it falls into the sea it can be traveling quite fast by the time it reaches the sea bed, especially in deepwater. Consequently, the dropped object can cause substantial damage to subsea equipment, with the potential for a serious environmental problem.

At the annual meeting of the Offshore Operators Committee (OOC) held in December 2010 it was noted that 19.5% of offshore accidents are due to crane movement (closely associated with dropped objects events) and that the number of accidents of this type area has not been improving. For this reason the OOC, working with BOEMRE, is researching the causes of such accidents so that actions can be taken to reduce their number.

Helicopters

Helicopters are used to transport personnel and light freight to and from offshore platforms. They are also used for the emergency evacuation of injured personnel (but cannot be used, of course, if the platform itself is sinking or on fire). The crash of a helicopter is almost always a very serious event, often leading to fatalities and serious economic loss.

Ship collisions

Offshore platforms can be hit by ships—usually the service boats that provide equipment and supplies. Sometimes these impacts can be very serious. In the case of the Mumbai High incident, for example, a large support vessel approached the platform to evacuate an injured man. The boat had problems with its computer-assisted azimuth thrusters so she was brought in stern-first under manual control.

The helideck on the support vessel hit a riser, which started leaking. The leak resulted in a fire that lead to approximately 22 fatalities and total loss of the platform and of the vessel.

REFERENCES

Broadribb, M. April 2008. 3 years on from Texas City. 4th Global Congress on Process Safety.

Bromwich, M. May 2011. Future of Offshore Oil and Gas Development in the U.S. Offshore Technology Conference.

Fewtrell, L., Bartram, J. (eds.) Water Quality Guidelines, Standards and Health. London: IWA World Health Organization.

National Commission on the BP Deepwater Horizon Oil Spill and Offshore Drilling. January 2011.

Pitblado, R. October 2008. Global process industry initiatives to reduce major accident hazards. 11th Annual Symposium. Mary Kay O'Connor Process Safety Center.

Sutton, Ian. 2010a. Behavior-based Safety. www.stb07.com/occupational-safety/behavior-based-safety.html.

CHAPTER

Major events

2

CHAPTER CONTENTS

INTRODUCTION

In principle, companies implement safety management systems because it is the right thing to do: they save lives, protect the environment, and reduce losses of equipment and production. In fact, of course, it is human nature to respond to events after they have happened. The closing of stable doors invariably follows the bolting of horses. So it is with the offshore oil and gas industry. Safety management systems, and the matching regulations, are developed in the aftermath of catastrophic events. And companies tend to be more thorough and rigorous in the application of existing systems in the months following a catastrophe.

This chapter describes eight of the most important events that have occurred in the process industries over the last 40 years. Three of the incidents occurred onshore but their impact was so profound that they affected the entire energy-related business—regardless of location or industry type. (One of the offshore incidents—Blackbeard—was actually a nonevent, but it offers profound lessons to managers and leaders at all levels.)

The importance of these events is not just to do with the loss of life and the environmental damage that was caused, but also in the lessons that can be learned from

Ian Sutton: Offshore Safety Management. DOI: 10.1016/B978-1-4377-3524-6.00002-2
Copyright © 2012 Ian Sutton. Published by Elsevier Inc. All rights reserved.

them. In particular, this chapter examines the lessons that can be learned by those designing and running offshore safety management systems.

The selected events are summarized in Table 2.1, which has seven columns:

(1) The name of the event

(2) The year in which it occurred

(3) Its location

(4) The type of operation (drilling, transportation, production—offshore or onshore)

(5) A very brief description as to what happened

(6) Its environmental and economic impact

(7) The number of fatalities

The events summarized in Table 2.1 are described in the remainder of this chapter. The purpose of the descriptions is merely to provide enough information to put the events in the context of what went wrong with the facility's safety management system. More detail is, of course, available in the one or more voluminous investigation reports written for each incident.

For each event the following information is provided:

- A description of the event itself
- A brief technical analysis
- The long-term impact
- Lessons learned, particularly for the offshore oil and gas industry
- The affected elements of safety and environmental management systems (SEMS)

The last of these—elements of SEMS—is to do with the SEMS rule that was introduced in 2010 for companies operating in the USA. (The rule is described in detail in Chapters 4 and 5.) The SEMS rule is built around 12 management and technical elements, as shown in Table 2.2. Those elements that seem most relevant to the event in question are highlighted and discussed.

SANTA BARBARA (1969)

In the late 1960s, the Union Oil Company of California, which later became Unocal and is now part of Chevron, was drilling about 6 miles off the coast of Summerland near Santa Barbara, California, in 188 ft (57 m) of water. One of the wells associated with the Alpha platform suffered a blowout on January 28, 1969. It took 10 days before the leak could be plugged with cement slurry.

It is estimated that up to a hundred thousand barrels of heavy oil was spilled into the Santa Barbara Channel. The slick it created covered 800 square miles. Many birds and other forms of wildlife died. There were, however, no human injuries or fatalities.

Table 2.1 Major Offshore Oil and Gas Events

Event	Year	Location	Operation Type	Description	Environmental and Economic Impact	Fatalities
Santa Barbara	1969	California	Drilling	Subsea blowout.	Extensive fouling of the coastline with crude oil. This incident was a major factor in the development of the modern environmental movement.	0
Flixborough	1974	England	Petro-chemical plant	Release of large quantity of highly flammable and explosive chemical (cyclohexane).	Massive explosion leading to many fatalities and total destruction of the facility.	28
Three Mile Island	1979	Harrisburg, PA, USA	Nuclear power plant	Partial meltdown of reactor core.	Very slight. There is on-going debate as to the long-term health effects of this event.	0
Piper Alpha	1988	North Sea	Production	Explosion and fire.	Total loss of the platform.	167
Valdez	1989	Alaska	Trans-portation	Release of crude oil from a grounded oil tanker.	Fouling of the shore line, leading to extensive loss of wildlife.	0
Blackbeard	2006	Gulf of Mexico	Drilling	Blowout (averted)	None	0
Montara	2009	Australia	Drilling	Blowout	Substantial spill of oil and extensive fire damage to a relief drill rig.	0
Deepwater Horizon	2010	Gulf of Mexico	Drilling	Release of gas and oil during drilling of a deepwater well.	Major environmental damage. Loss of a world scale drilling rig, enormous penalties and clean-up costs.	11
Fukushima Dai-ichi	2011	Northern Japan	Nuclear power plant	Severe damage to at least three nuclear power plants, with likely partial meltdown in at least one case.	Large swathes of land and parts of the ocean have been contaminated by an unknown amount of radioactivity. The economic consequences are enormous, not only for the nation of Japan, but the remainder of the world.	None at the time of writing, but this could change.

Table 2.2 Elements of SEMS

(1)	Safety and Environmental Information
(2)	Hazards Analysis
(3)	Operating Procedures
(4)	Training
(5)	Pre-Startup Review
(6)	Assurance of Quality and Mechanical Integrity of Equipment
(7)	Safe Work Practices
(8)	Management of Change
(9)	Investigation of Incidents
(10)	Emergency Response and Control
(11)	Audit of Safety and Environmental Management Program Elements
(12)	Records and Documentation

The event

The following is a brief summary of the sequence of events:

- The company had received a waiver to use lower quality casing than required by regulation.
- They had drilled to 3500 ft below the mud line.
- As riggers were retrieving pipe in order to replace a drill bit, the amount of mud available became very low.
- A natural gas blowout occurred.
- The hole was capped, but this led to a large build up of pressure that caused five breaks in an east—west fault on the ocean floor. Oil and gas spilled from these breaks, some of which were miles away from the original blowout.
- It took 11 days before the incident was brought under control.
- Another leak occurred some weeks later.

Incident analysis

The proximate cause of the event was that the drill bore casing was below federal and California standards. Unocal had been given permission to use the lower strength casing by the regulator: the US Geological Survey. (The rig was more than 3 miles from the coast, so the State of California's more rigorous standards did not apply.)

A root cause for the event was that well-established rules and procedures were not followed.

Long-term impact

Other environmental catastrophes in the USA occurred at about the same time as the Santa Barbara spill. The most famous of these was probably when the highly-polluted Cuyahoga River in Ohio burst into flame about 6 months later. There was also a heightened general awareness of air and water pollution in general, and of the impact of toxic chemicals such as DDT.

In response to the public outcry that followed the Santa Barbara incident and to other high-profile events that occurred at about the same time, the United States Congress passed the National Environmental Policy Act in 1969. The Act required federal agencies to file impact statements for all actions that could have a significant ecological effect. Within two years President Richard Nixon had created the Environmental Protection Agency and signed into law the Clean Air Act. National Earth Day was started the following year; Sierra club membership doubled during this period.

The incident also showed that management at that time was not really attuned to environmental and safety issues, as illustrated in the quotation below.

> *I don't like to call it a disaster because there has been no loss of human life. I am amazed at the publicity for the loss of a few birds.*
>
> *Fred L. Hartley, President of Union Oil Co.*

Lessons learned

Three important lessons come out of the Santa Barbara incident:

(1) Follow the rules

(2) Understand the importance of public response

(3) Evaluate the economic impact

Follow the rules

The first and most obvious lesson learned is that rules and standards are generally sound and they should be followed. Management must have very good reasons for circumventing those rules, and those in charge of daily operations need to know what the rules are, and to make sure that they are followed. (This simple failure to follow the rules is a factor in a number of the other events discussed in this chapter.)

Public response

The second lesson is that offshore accidents can generate very strong and negative response, even if the direct impact on the public is not that large. If an event is serious enough, it will result in enormous publicity, new regulations, and high costs associated with clean up and litigation (something that is being relearned in the wake of the Deepwater Horizon (DWH) incident).

Economic loss

The third lesson from the Santa Barbara spill is that events such as this can have enormous financial consequences (once more, a lesson being relearned following DWH).

Table 2.3 Elements of SEMS

(1)	Safety and Environmental Information
(2)	Hazards Analysis
(3)	***Operating Procedures***
(4)	Training
(5)	Pre-Startup Review
(6)	Assurance of Quality and Mechanical Integrity of Equipment
(7)	***Safe Work Practices***
(8)	Management of Change
(9)	Investigation of Incidents
(10)	Emergency Response and Control
(11)	Audit of Safety and Environmental Management Program Elements
(12)	Records and Documentation

Since the time of this event drilling off the coast of California has been severely restricted in spite of the fact that the additional revenues generated would provide welcome relief to that state's stretched finances.

Elements of SEMS

The elements of SEMS (the SEMS rule) that appear to be particularly relevant to the Santa Barbara incident are shown in Table 2.3.

Operating procedures

The accident happened a long time ago, and it is not known now if the drill crew had been provided with good operating procedures. However, the fact that they did not follow correct procedures as prescribed by law suggests that either the procedures had not been written or they were not followed.

Safe work practices

The comments made above regarding operating procedures can also be applied to the topic of safe work practices. The drill crew chose to override the practices that they were supposed to follow, resulting in a blowout.

FLIXBOROUGH (1974)

The Flixborough event is one of the three incidents in this chapter that occurred onshore. However, this event is extremely important to all types of process industry because it led to the creation of process safety management (PSM) systems—which in turn provided the foundation for many other similar systems, including safety and environmental management program (SEMP) and SEMS.

The Flixborough chemical plant produced caprolactam, which is in turn used to manufacture nylon. The process used six large pressurized reactors containing cyclohexane. (Cyclohexane is comparable to gasoline; it is a liquid at ambient conditions, but vaporizes easily and is highly flammable.) The plant was located on the east coast of England; it had been in operation since the year 1967 and was operated by Nypro (a joint venture between Dutch State Mines and the British National Coal Board).

The event

Prior to the actual accident, which occurred in March 1974, a vertical crack had appeared in Reactor #5. It was decided to remove this reactor and to install a bypass, connecting Reactors 4 and 6. The bypass pipe had been designed by engineers who were not experienced in high-pressure pipework, had no plans or calculations, it was not pressure-tested, and was mounted on temporary scaffolding poles that allowed the pipework to twist under pressure. Moreover, the bypass pipe was of a smaller diameter (20 inch) than the reactor flanges (24 inch) so, in order to align the flanges, short sections of steel bellows were added at each end of the bypass. Due to the dog-leg shape of the bypass these bellows probably squirmed when under pressure.

During the late afternoon on June 1, 1974, the 20-inch bypass failed catastrophically due to lateral stresses in the pipe, probably during a pressure surge (which may have been caused by a fire at a nearby 8-inch pipe). The rupture resulted in the release of around 40 tons of cyclohexane to the atmosphere. The cyclohexane/air mixture and found a source of ignition and caused a massive explosion.

All 18 persons in the control room died as a result of the windows shattering and the collapse of the roof. Overall, there were 28 deaths and 89 injuries (both workers and public).

Incident analysis

The proximate cause of the incident was the lack of proper engineering and failure to implement Management of Change (MOC).

However, one of the recurring issues in the investigation of large accidents is that the causes are often more complex and subtle than originally thought. For example, in the case of the Flixborough event, one of the original investigators, scientist Ralph King, believes that the cause was more than a simple mechanical failure. Instead, he believes that the event was triggered by a process upset caused by the addition of water to the reactors and a failure of the agitation system.

Long-term impact

The Flixborough event was seminal; nothing in the process industries in the United Kingdom was the same afterwards. The event greatly increased public concern over industrial plant safety, leading to increased regulations in the United Kingdom regarding hazardous industrial processes (the COMAH regulations are used now).

Although the Flixborough accident was very tragic, it did lead to some positive developments. In particular, as discussed above, the onshore process industries developed new safety management systems that eventually became known as PSM. For example, the first paper on the topic of Hazard and Operability Study (HAZOPs) was presented by Lawley in the year 1974. And the accident almost single-handedly forced the development of MOC systems in the onshore process industries.

Lessons learned

Probably the most important lesson from Flixborough for the offshore industries is that it is necessary to have a Safety Management System. The Flixborough accident did not occur because one or two of the elements of PSM/SEMS were badly implemented. It occurred because Nypro did not have a Safety Management System and because there was inadequate oversight of the industry on the part of the government regulators.

The above situation still applies to some of the companies working offshore. For example, Table 5.4 in Chapter 5 shows that approximately half of the companies operating in the Gulf of Mexico (GoM) do not have a complete SEMP in place. This does not mean that they have nothing in the way of Safety Management Systems. Nevertheless it would appear as if there is substantial room for improvement.

Similarly, with regard to the regulations and regulators, the drastic changes that have occurred in the regulatory regime following the DWH incident indicate that what was present before, was not sufficient.

Elements of SEMS

Table 2.4 shows the elements of SEMP/SEMS that are of particular relevance to the Flixborough incident.

Hazards analysis

There is no indication of any sort that a Hazards Analysis had been carried out on the modified system prior to its being installed. Had such an analysis been carried out, it is likely that the analysis team would have called for more rigorous engineering on the bypass line and bellows. There would also have been discussion about control and emergency response systems.

Mechanical integrity

The Mechanical Integrity element of PSM was also a critical issue as the temporary pipe had not been designed according to established engineering standards. Also, there must have been problems with the integrity or inspection program before the accident. After all, if the leak in the pressure vessel had not occurred, there would have been no need to install a bypass.

Management of change

One of the most important lessons learned from Flixborough was to do with MOC. The pipe that failed was modified without an engineering analysis being carried out—the change was a classic case of what is sometimes referred to as "midnight engineering."

Table 2.4 Elements of SEMS

(1)	Safety and Environmental Information
(2)	***Hazards Analysis***
(3)	Operating Procedures
(4)	Training
(5)	Pre-Startup Review
(6)	***Assurance of Quality and Mechanical Integrity of Equipment***
(7)	Safe Work Practices
(8)	***Management of Change***
(9)	Investigation of Incidents
(10)	***Emergency Response and Control***
(11)	Audit of Safety and Environmental Management Program Elements
(12)	Records and Documentation

The topic of MOC became so closely identified with the Flixborough event that, at a process safety conference some years later, one session was called, "Management of Change—excluding Flixborough."

Emergency response and control

One of the saddest aspects of this incident was the fact that there was probably a considerable time gap between the start of the large leak and of the explosion itself. During that time, the operators were presumably trying to control the leak and to take other emergency control actions. Yet they should have realized that they had "lost it." They should have quickly departed from the scene and saved themselves. After all, although the explosion damaged nearby houses, no members of the public were killed. Therefore, had the operations personnel escaped the scene rather than trying to contain the leak, it is likely that they would have survived. The plant would have been destroyed—but that happened anyway.

In addition, as already noted, the control room was not blast-resistant. Everyone in it died. Had this building been hardened, their chance of survival would have been much greater.

THREE MILE ISLAND (1979)

The Three Mile Island incident—usually referred to by its initials of TMI—occurred at a nuclear power plant located near Harrisburg, Pennsylvania, in the year 1979. The island on which the facility is located gets its name from the fact that it is 3 miles long.

Figure 2.1 is a sketch of the TMI process, which uses pressurized water reactor technology.

The system is comprised of three water circuits. The first is pressurized water that flows through the reactor core. The water picks up heat from the nuclear fuel and its temperature rises. The pressure in this loop is high enough to prevent steam formation.

The water in the primary loop is pumped through the two steam generators. These are heat exchangers in which a secondary water stream flows through the shell side. Heat is transferred from the primary water to the secondary water—which is allowed to boil. The steam turns the turbines which generate the electricity that goes to the plant's customers. The steam leaving the turbines is condensed, "polished," and pumped back to the steam generators.

The third water circuit consists of cooling water that condenses the steam in the condenser. The cooling water is itself cooled in large natural draft cooling towers (the structures that are the most visible part of the plant when seen from a distance).

In an emergency the control systems will shut down the reactors by inserting control rods into the core (this is known as a "scram"). Unfortunately, this action does not immediately stop the nuclear reaction. Heat continues to be generated in the nuclear core even after shutdown, so it is essential that all three of the water systems continue to operate so that heat can continue to be removed for many hours. This need for on-going cooling is in contrast to most process plants in the hydrocarbon industry—a shutdown will immediately remove sources of heat, and temperatures will immediately start to fall to a safe value. Hence, the continued operation of cooling systems is not likely to be so critical (except maybe with respect to the high temperature firebrick on the inside of fired heaters).

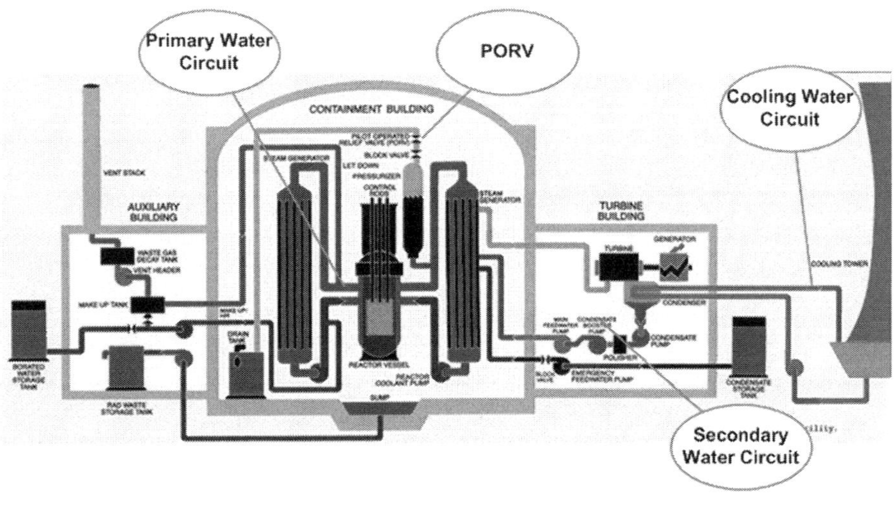

FIGURE 2.1

Pressurized Water Reactor Sketch.

Shown in Figure 2.1 is a vessel known as a pressurizer (it is the half-full vessel to the left of the right-hand steam generator). The pressurizer is a part of the primary water circuit, and plays a critical role in controlling the pressure and temperature in that circuit. If the reactor temperature rises, then the volume of water increases and the level in the pressurizer increases (and vice versa). If the pressure becomes too high, the Pilot Operated Relief Valve (PORV) opens and steam is vented to the drain tank.

From a risk management point of view, the most obvious weakness of the system shown in Figure 2.1 is that each of these water circulation loops must work for the overall system to work. And, since each system relies on the proper operation of pumps and associated equipment, the system is not inherently safe. Should the flow of water in any of the three loops stop, the reactor heat will not be removed. This will cause the reactor temperature to rise. If this temperature rise is not controlled, there may eventually be a "meltdown" in which the core structure melts through the floor of the containment building and enters the ground, where it could cause an explosion if and when it comes into contact with subsurface water. Alternatively, the high temperature in the reactor leads to the formation of hydrogen and other gases that can cause an explosion.

An important safeguard is that the reactor is located inside a containment building, a pressurized structure whose purpose is to prevent any leaks escaping to the atmosphere.

The event

The TMI event resulted from a long sequence of human, equipment, and instrument failures. A very brief overview of what took place is described below. There are, of course, many reports and analyses that describe this complex event in great detail.

- Operators were trying to remove trash from the secondary water system.
- They used instrument air to blow the system clear. This is an unacceptable practice as any reverse flow could contaminate the instrument air and cause a myriad of control problems, which is what happened. Water entered the instrument air system through a failed check valve.
- The contamination in the instrument air led to control problems that caused the main feed water pumps in the secondary circuit to stop.
- The backup secondary water pumps were blocked in for maintenance. This was a major rule violation—before these pumps were removed from service the reactor should have been shut down in a controlled manner.
- The reactor pressure rose.
- Reactor rods automatically inserted themselves into the reactor core, leading to a shutdown of the reactor. Once this happens the nuclear reaction starts to slow down, but, as discussed above, it does not come to an immediate stop. Heat continues to be generated in the core.

- Because heat was still being generated, the PORV, shown in Figure 2.1, opened in order to reduce the pressure.
- The pressure in the primary water circuit fell, at which point the PORV should have closed. In fact, it stuck open. However, this was not known to the operators. Their only indication was that the signal to the valve was for it to close. They did not have a direct measurement of the actual valve status.
- Large quantities of water and steam flowed through the open PORV to the drain tank.
- The pressure in the primary water circuit fell to the point where the water in the pressurizer started to boil. The boiling water gave a false indication that the vessel was full of water. The level indicator in the pressurizer was reading incorrectly; it showed high level even when the water level was in fact low.
- The operators responded to the apparent filling of the vessel by turning off the quench pumps (they were trained not to let the pressurizer "go solid"). They had lost faith in all alarms, and did not respond to them.
- Their actions further reduced core cooling and increased steam formation.
- The fuel rods became partially exposed.
- The drain tank filled. Its rupture disk burst, thus allowing radioactive water and steam into the containment building. Radiation levels inside the building were 300 times the normal value.
- The primary water pumps started to cavitate—a sign of low level in the pressurizer. However, the operators did not recognize what was going on—they still believed that they were facing a high-level situation.
- At shift change, one of the new operators recognized what had happened, and finally closed the PORV. However, they still did not add fresh water to the circuit. (Had they done so, it is likely that the cooling would not have been very effective because the fuel elements had fused together.)
- Meanwhile the zirconium cladding on the outside of the reactor rods, which was now at a much higher than normal temperature, reacted with the steam to form hydrogen. This led to an explosion which allowed for some radioactive gases to collect in the containment building.
- Some of these gases escaped to the atmosphere.

Eventually, the situation was brought under control, and only a minimal amount of radiation was released to the atmosphere. Nevertheless, inspections of the reactor core in subsequent years showed that the damage was extensive, and that they had been nearer to a meltdown situation than they realized at the time.

The following additional observations can be made about the incident:

- The initial cause of the event—contamination of the instrument air leading to water pump shutdown—was seemingly a very minor matter.
- There had been serious rule violations (using instrument air to flush a line, and continuing operation while the backup pumps were out of service).

- The operators were trapped in a set of assumptions as to what was going on—even though there were plenty of indications that those assumptions were incorrect.
- Once a new group of operators became involved they quickly recognized what was going on. They were not trapped by the previous way of thinking.
- There were so many alarms going off that the operators simply ignored them. This phenomenon later came to be known as "alarm flooding."

Incident analysis

Discussion of TMI in the literature will often state that the incident was a major factor in the creation of the modern human factors discipline. This is certainly true, but may be somewhat limiting if human factors are considered to be confined to issues such as ergonomics and the man–machine interface. In particular, the failure of senior operators, supervisors, and managers to recognize what was going on (and to recognize that some instruments were giving them the wrong signal) can be treated as a different form of human factors.

Long-term impact

The long-term consequences of TMI were profound. Although no one was injured or killed, and although it appears as if public health was not much affected, the event led to a moratorium on the construction of new nuclear power plants in the USA. This moratorium was caused not only by the direct public relations impact of TMI but also by the increased capital costs, some of which were a result of the more stringent regulations and standards that followed.

By the year 2010, the impact of the event had diminished somewhat, and, given the increased cost of conventional fuels, something of a "nuclear renaissance" was under way. However, it seems likely that the Fukushima Dai-ichi accident (see p. 71) will nip this renaissance in the bud.

Lessons learned

The lessons outlined above regarding human factors, troubleshooting, fixation and following the rules all apply to the offshore oil and gas industry. However, the biggest lesson could simply be that so much can be learned from an industry that appears, at first sight, to be so different from an offshore oil and gas platform.

Professionals in all industries have a tendency to say, "But we're special, we're different from all those others." And, of course, there are profound differences in technology between industries. An offshore platform does not have to worry about a nuclear core that must be cooled, even when shut down. In return, the operator of a nuclear power plant does not need to know how to drill a well with amazing precision a mile below the surface of the sea in total darkness.

Yet, with respect to safety management systems, these differences are not nearly as profound as the similarities, as even the most superficial examination of the TMI shows. Indeed, as discussed in Chapter 4, the offshore oil and gas industry would appear to be following some of the approaches used by the nuclear power industry—specifically with regard to the creation of the Institute of Nuclear Power Operations (INPO) and the API's Center for Offshore Safety.

Follow the rules

First and foremost, the incident would never have occurred if the operations and maintenance personnel had simply followed well-known and well-established rules. Violations included the use of instrument air to blow out a line, and taking critical backup equipment out of service while the facility was still operating. None of these actions were necessary—presumably they were taken because doing so saved a small amount of time and trouble.

Troubleshooting

One of the features of this event is that, unlike Fukushima Dai-ichi, there was no significant damage to equipment or instrumentation. At any time, had management and the operating personnel understood what was going on, they could have taken corrective action. All systems were available to them. The limitations and incorrect actions existed only in the minds of the managers and operators—there were no external events such as tsunamis or earthquakes to consider. Had their troubleshooting skills—as distinct from emergency response training—been better, they would have figured out what was happening, and taken the corrective actions quickly. In its initial stages, the TMI incident was not an emergency. The system was in trouble, but no one's safety was threatened.

A troubleshooting response is appropriate when there is no anticipated danger to personnel and when there is little chance of any significant equipment damage; usually the main concerns are about product quality, production rates, productivity, and equipment repair costs. Examples of "trouble" include:

- Product quality problems
- Erratic machinery performance
- Preparation for extreme weather conditions such as hurricanes or ice storms
- Minor environmental problems
- Reduced yields of raw materials and/or increased energy consumption

Fixation

Both management and supervision suffered from the problem of "fixation." Although they had many sources of information telling them that the water level in the reactor was low, they elected to believe the (incorrect) instrument that told them that the level was high.

Fixation of this type is a common phenomenon when people are overloaded with information, some of it conflicting. They screen out any information that does not conform to the understanding that they have already established.

Critical information missing

In order to properly control a nuclear reactor, it is critical that the operators know what the level of water in the reactor is. At TMI a direct water level measurement was not available; the operators relied on secondary information, particularly the level of water in the pressurizer, to tell them how much water was in the reactor.

Operator–instrumentation interface

The operators had trouble finding the right information in order to determine what was going on. Issues included the following:

- Some control panel readings were obscured by pieces of paper.
- There were too many alarms, and the alarm printer jammed up. The operators could not make sense of it all.
- A control signal (to the PORV) was interpreted as giving the actual status of the valve. In fact, the PORV had not closed.

Inherent safety

The TMI power plant relied on three separate cooling water systems to work properly. Each of those systems had control valves and rotating equipment that had to work properly. None of these loops was inherently safe. There were many backup systems, but these were not effective, even though, unlike Fukushima, all of the equipment was available had the operators known what was going on.

This system was definitely not inherently safe, which means that, sooner or later, a sufficient number of things will go wrong, resulting in an overall system failure.

Elements of SEMS

Nearly all of the elements of SEMS played a role in TMI. Table 2.5 lists some of the more prominent of these.

Table 2.5 Elements of SEMS

(1)	***Safety and Environmental Information***
(2)	Hazards Analysis
(3)	Operating Procedures
(4)	Training
(5)	Pre-Startup Review
(6)	Assurance of Quality and Mechanical Integrity of Equipment
(7)	***Safe Work Practices***
(8)	Management of Change
(9)	Investigation of Incidents
(10)	***Emergency Response and Control***
(11)	Audit of Safety and Environmental Management Program Elements
(12)	Records and Documentation

Safety and environmental information

The operating team at TMI were not provided with the real-time critical information that they needed (such as the actual level in the reactor), and some of the information that they did receive (such as the status of the PROV) was erroneous.

Safe work practices

Failure to follow good safe practices was critical. The operators should not have used instrument air to clear a line, and management should not have authorized the disabling of a spare pump while the facility was still in operation.

More broadly, the difficulties that the operators experienced with the man–machine interface showed that safe work practices require a good understanding of human factors issues (even though this topic is not explicitly identified in Table 2.5 as being one of the elements of a SEMS program).

Emergency response and control

The emergency response and control plan was clearly inadequate, for many of the reasons discussed above (such as the operators being confused as to the true water level in the pressurizer). It is important not only to have emergency plans, but also to develop troubleshooting procedures that outline actions to be taken during the initial stages of a crisis.

PIPER ALPHA (1988)

The Piper Alpha platform was a large, fixed-structure platform located about 110 miles north east of Aberdeen in 474 ft of water. It handled both oil and gas. Equipment on board separated the incoming oil and gas streams and created (hydrocarbon) condensate product. The oil and condensate were exported from Piper Alpha through a 30-inch line to the Flotta Terminal. Piper Alpha contributed about 10% of the oil production from the UK sector of the North Sea.

The explosion and fire that occurred in 1988 is one of the most important process safety events that have ever occurred—either onshore or offshore.

Critically important to an understanding of this event is the fact that Piper Alpha was a hub platform. Not only did the platform have its own risers, it also received gas from the Tartan "A" platform, which was located about 11.5 miles away. The gas flowed to Piper Alpha through a subsea 18-inch pipeline. This gas stream was combined with Piper's own gas, compressed and sent to the MCP-01 platform, some 33.5 miles away. Some of the gas was also used as lift gas and as fuel for the onboard generators (the gas could also be flared). In addition, another platform—Claymore—could also feed gas to Piper Alpha. All of the platforms were operated by the same company: Occidental.

The event

The complex sequence of events leading up to the explosion and fire is described in detail in Appendix 19 of *Lees' Loss Prevention in the Process Industries* (2004). A brief overview is provided below.

At 10:00 p.m. on July 6, 1988, a massive explosion followed by a large fire led to the destruction of the platform. One hundred and sixty-seven men died (62 escaped). Because the accident occurred in the evening, many of the persons on board were off duty and were located in the living quarters. The smoke from the fire trapped these persons, accounting for most of the fatalities.

The platform had two hydrocarbon condensate pumps: "A" and "B." Either pump had the capacity to handle the total condensate flow—the other pump was a 100% spare. Prior to the accident the "A" pump was down so that three maintenance tasks could be carried out on it. They were:

- A full 24-month preventive maintenance
- Recertification of the pump's relief valve
- Repair of the pump coupling

The pump had been shut down for the first task: preventive maintenance work. However, it was isolated only by closed gas-operated valves on the suction and delivery lines. It had not been positively isolated, i.e. blinds (slip plates) had not been installed on those lines.

The relief valve had been removed in the morning for testing (the second of the above tasks). Two blind flanges were put on the pump casing at the location of the relief valve. The flanges were not bolted down tightly.

At the evening shift changeover it was decided not to carry out the preventive maintenance work after all. Therefore, the only work to be done would the third task, the repair of the coupling—which did not require access to the pump itself, and so did not require that the pump be positively isolated. Due to a failure in communications, information as to the status of the open work order to do with the relief valve had not been properly communicated to the incoming shift. In particular, the incoming crew did not know that the relief valve blind flanges had not been tightened up.

At 9:50 p.m. the "B" pump tripped out and could not be restarted. If neither of the two condensate pumps could be started within 30 minutes then the platform would have to be shut down—an expensive decision—so the decision was made to put pump "A" back into service.

Shortly after the "A" pump was started condensate leaked from the loose relief valve flanges. The condensate found a source of ignition that led to an explosion that caused extensive damage and that probably led to the deaths of seven men adjacent to the condensate pumps. The explosion was followed by an oil pool fire and a blowdown of gas inventories to flare. The oil fire created a massive smoke plume that made the escape routes from the living quarters (where most of the personnel were located) impassable.

About 20 minutes after the initial explosion the gas riser from the Tartan platform, which had not been shut in, failed and a second massive explosion occurred, followed by an escalation of the fire. Fifty minutes later the riser going to MCP-01 also failed, resulting in a third explosion. A continued series of explosions and fires led to the eventual destruction of the platform and to the deaths of many additional workers.

Incident analysis

A review of the causes of the accident shows that many, many factors came into play—there was a broad, systemic failure in the safety management system. Some of these causes included the following:

- The platform had firewalls. However, they were not blast-resistant, so they failed when subject to explosions, thus preventing neither the effects of explosion or fire.
- The Emergency Shutdown System (ESD) automatically shut off the flow of oil from the platform to Flotta, but it did not stop the incoming flow of gas on the grounds that doing so would shut down other platforms.
- The main communications for the complex of platforms were located on Piper. The explosions took out most of those communications, thus degrading the emergency response.
- Emergency escape routes were blocked.
- The firewater system failed to operate properly.

Long-term impact

The long-term effects of the Piper Alpha event were profound. The formal investigation into the Piper Alpha tragedy was headed by Lord Cullen, a Scottish High Court judge. In the year 1990 his committee published the *The Public Inquiry into the Piper Alpha Disaster*, a document that is usually referred to simply as "The Cullen Report."

The two dominant offshore cultures at the time—the North Sea and the GoM—each reacted to the report's findings vigorously, but those responses were different from one another. Companies operating in the North Sea had already put in place a Safety Case culture (which included Formal Safety Assessments). However, the Cullen report was very critical of the manner in which many of the safety cases had been prepared and implemented. In response to this criticism, companies operating in the North Sea (and later in other areas such as Australia and Southeast Asia) made great strides in improving the quality and implementation of their safety cases. (Chapter 6 discusses the Safety Case technique.)

Companies operating in the GoM had an equally forceful response to the Piper Alpha incident, but their approach was strategically different. There had not been a Safety Case system prior to Piper Alpha; instead there were thousands of platforms (many of them were simple, small, and similar to one another). Rather than preparing individual safety cases for each of these near-identical platforms, it was felt to be more appropriate to create a set of relatively prescriptive standards which everyone would follow. Most of these standards were developed by the American Petroleum Institute (API), and fall under the overall direction of the API RP 75 which called for companies to develop a SEMP as discussed in the next chapter.

Neither of the two approaches—Safety Cases or the use of API standards—is inherently better than the other; each was designed to fit the circumstances of the industries to which it applied. Moreover, there is considerable overlap between the two approaches. For example, many safety cases make reference to API RP 14C because that standard provides sensible and practical guidance to do with the analysis of hazards and the selection of instrument systems.

Moreover, the two approaches appear to be moving toward one another. In the North Sea declining production has resulted in small companies taking over platforms from industry majors. These new owners do not have the financial depth to prepare elaborate safety cases—instead they simply want to be told what the rules are, and what they are expected to do, just like the smaller operators in the GoM.

In the GoM, on the other hand, the trend has been to deep water, high capacity platforms. These platforms are expensive. Hence, there is a tendency for the operators of these platforms to develop safety cases—often under a different name— to limit their financial risk.

Lessons learned

Piper Alpha generated a huge number of lessons learned, many of which provide the basis for the content of this book. Aside from the development of new safety standards and calculation techniques and the writing of new regulations, one of the most important lessons was to do with intervention. The inventory of gas and oil on a platform such as Piper Alpha is quite low, and an emergency always generates a system blowdown. Had the flow of gas from the other platforms been stopped at once, it is likely that the fire would have burned itself out and the number of fatalities would have been limited to those caused by the initial explosion.

Another important lesson learned from this event is to do with the lack of leadership, particularly with regard to the men who were left in the living quarters. No one on the platform showed the leadership needed to move people to a location of the platform where they had a better chance of escaping.

Further discussion to do with the topic of leadership, and its absence in this incident, is provided in Chapter 8.

Elements of SEMS

Nearly all of the elements of process safety/SEMS played a role in the Piper Alpha incident—so profound was its nature. Four elements, in particular, are identified in Table 2.6.

Pre-startup review

Some major engineering and equipment changes were made to the platform in the day before the event. Before equipment is started, the system should be checked with a Pre-startup Safety Review to make sure that all changes have been properly managed and the system is safe to operate.

Safe work practices

The failure of work practices—particularly the handover of work orders from one shift to another—was a major contributing factor.

Table 2.6 Elements of SEMS	
(1)	Safety and Environmental Information
(2)	Hazards Analysis
(3)	Operating Procedures
(4)	Training
(5)	***Pre-Startup Review***
(6)	Assurance of Quality and Mechanical Integrity of Equipment
(7)	***Safe Work Practices***
(8)	Management of Change
(9)	***Investigation of Incidents***
(10)	Emergency Response and Control
(11)	Audit of Safety and Environmental Management Program Elements
(12)	***Records and Documentation***

Investigation of incidents

The investigation of this incident resulted in the highly influential Cullen report, which in turn led to many profound changes in the way that offshore safety is managed. Therefore, it can be said that this topic was handled extremely well.

Records and documentation

The lack of proper records and documentation, particularly with regard to the information to do with changes that were made to the compressor valves, played a large role in this event. Another postaccident issue was that many of the records—including anything written about the work permits—were destroyed in the fire.

VALDEZ (1989)

On March 24, 1989, the oil tanker *Exxon Valdez* ran aground near the port of Valdez in Alaska. Eight of the ship's cargo tanks were ruptured and approximately 250,000 barrels of oil were released into the ocean. The accident had major environmental and long-term economic consequences but no one was injured. Spilled oil eventually covered some 1300 miles (2100 km) of coastline.

The event

Exxon Valdez left the Valdez oil terminal in Alaska on March 23, 1989; she was carrying a full load of crude oil, and was bound for Long Beach, California. The outbound shipping lane was obstructed with icebergs, so permission was obtained from the Coast

Guard to go out through the inbound lane. The ship was on autopilot when she struck Bligh Reef at 12:04 a.m. on March 24, 1989. A critical piece of navigation equipment, the sonar, had been out of service for many months. Had it been in operation it is likely that the submerged reef would have been observed and avoided.

It was widely reported that the captain of the ship was intoxicated. Although he had been drinking before the ship sailed, at the time of the accident he was asleep in his cabin. Therefore, his condition did not contribute directly to the event, except that he allowed subordinate officers to be in charge of the ship's fate.

Long-term impact

In response to the spill, Congress passed the Oil Pollution Act of 1990 (OPA). The legislation included a clause that prohibits any vessel that, after March 22, 1989, has caused an oil spill of more than one million US gallons (3800 m^3) in any marine area, from operating in Prince William Sound.

The OPA also set a schedule for the gradual phase-in of a double hull design, providing an additional layer between the oil tanks and the ocean. While a double hull would likely not have prevented the Valdez disaster (in fact, some of the leaking tanks were already within a double hull structure), a Coast Guard study estimated that it would have cut the amount of oil spilled by 60%. All tankers traveling through the same area are scheduled to be double-hulled by 2015.

Lessons learned

One of the most important lessons learned was the reaction of the ExxonMobil company. Their management recognized that the causes went well beyond a single officer who had had too much to drink, or failure to maintain critical instrumentation. A fundamental rethink of safety management systems was needed. The company did go through this rethinking process, and the results can be seen with respect to their response to the Blackbeard event, as discussed in the next section.

Elements of SEMS

Table 2.7 shows the elements of SEMP/SEMS that are of particular relevance to this incident.

Pre-startup review

The *Exxon Valdez* had critical safety equipment (the sonar) that was not working properly. If a pre-sail-away review had been properly carried out, this deficiency would have been recorded and corrected. At the very least, some type of hazards analysis would have helped determine if the ship was safe to sail. Also, the identification of such a deficiency would have led to the issuance of temporary operating procedures, possibly the selection of a safer route, and training for those involved.

Safe work practices

The company's safe work practices should have incorporated procedures for what to do when events occurred such as failed instruments or the captain being unavailable.

Table 2.7 Elements of SEMS

(1)	Safety and Environmental Information
(2)	Hazards Analysis
(3)	Operating Procedures
(4)	Training
(5)	***Pre-Startup Review***
(6)	Assurance of Quality and Mechanical Integrity of Equipment
(7)	***Safe Work Practices***
(8)	Management of Change
(9)	Investigation of Incidents
(10)	Emergency Response and Control
(11)	Audit of Safety and Environmental Management Program Elements
(12)	Records and Documentation

BLACKBEARD (2006)

Because the Blackbeard incident was actually a nonevent, no official reports or analyses are publicly available. Therefore, most of the information to do with the event is based on news stories and information that ExxonMobil chose to share with the media, particularly an interview that the company's Chief Executive, Rex Tillerson, gave to the *New York Times*.

In February 2005, the oil company Exxon started drilling the Blackbeard West well, 28 miles off the coast of Louisiana, and about 70 miles from the site of the future DWH event.

The (non)-event

At the time of the event Blackbeard was not only in deep water but it was also the world's deepest oil well: 32,000 ft below the seabed. Company experts reported that the well had potentially had huge reserves—up to 1 billion barrels. However, the prospect was also risky because of the very high temperatures and pressures in the well.

By September 2006, 500 days after the start of drilling, the well had reached a depth of 30,067 ft, a record at the time, and was within about 2000 ft of its target. However, well conditions were described as being "hellish"—the drillers were experiencing very high temperatures and pressures (more than 29,000 psi). Indeed, the well had already experienced a kick. They were concerned that the well might blowout.

> *In the ensuing discussions as to whether to keep going or not, the chairman and CEO of Exxon, Rex Tillerson, sided with the drillers, i.e. he decided to shut down the project and plug the well with cement. Exxon wrote off Blackbeard as a $187 million dry hole.*
>
> *New York Times, July 12, 2010*

Long-term impact

Before the DWH accident, the embodiment of a disastrous oil spill was the 1989 grounding of the *Exxon Valdez* in Prince William Sound off the coast of Alaska (p. 63). That accident, for which Exxon was found responsible, led to a profound rethinking of safety management at the company. Exxon developed a rigid system of rules for all its operations, from gas stations to offshore platforms, and it empowered everyone, even contractors, to speak up about safety problems.

Lessons learned

The Blackbeard incident shows how a company culture can change. In the case of the Exxon, the Valdez event led to management instituting a safety culture that is among the best in industry.

Elements of SEMS

Table 2.8 shows the elements of SEMP/SEMS that are of particular relevance to this nonincident. But the feature of this event that is most outstanding is the leadership that

Table 2.8 Elements of SEMS

(1)	***Safety and Environmental Information***
(2)	Hazards Analysis
(3)	Operating Procedures
(4)	Training
(5)	Pre-Startup Review
(6)	Assurance of Quality and Mechanical Integrity of Equipment
(7)	***Safe Work Practices***
(8)	Management of Change
(9)	Investigation of Incidents
(10)	Emergency Response and Control
(11)	Audit of Safety and Environmental Management Program Elements
(12)	Records and Documentation

was shown, and the willingness to make difficult decisions. Some further thoughts on these topics are provided in Chapter 8.

Safety and environmental information

Based on available information, it seems apparent that details on this nonevent was available in a timely manner, and that the information was transmitted to senior management quickly. Hence, the incident could be controlled before it got out of hand.

Safe work practices

The incident showed the correct use of safe work practices at the senior management level. Although it can be assumed that there were no specific written instructions for the management of an event such as this, it is clear that management at all levels understood the importance of "doing the right thing."

MONTARA (2009)

In 2009, the Montara Wellhead Platform (WHP) suffered a blowout that led to the release of a large oil slick (there were no injuries). The spill continued for 74 days. It took five attempts to kill the well. It has been estimated that the total leak rate was on the order of 2000 barrels per day. During the recovery operations the West Atlas relief rig caught fire and was severely damaged (again, there were no injuries).

The event

The official report to do with the event—the Montara Commission of Inquiry (Montara 2010)—was published in June 2010, but only released to the public in November of 2010. The following summary of what took place is from the Commission's report.

> *In the early hours of 21 August 2009, a small "burp" of oil and gas was reported as having escaped from the H1 Well at the Montara WHP. The oil and gas had traveled a distance of over four kilometers from the reservoir beneath the sea bed. Whilst the initial 'burp' subsided, approximately two hours later the H1 Well kicked with such force that a column of oil, fluid and gas was expelled from the top of the well, through the hatch on the top deck of the WHP, hitting the underside of the West Atlas drilling rig and cascading into the sea.*
>
> *For a period of just over 10 weeks, oil and gas continued to flow unabated into the Timor Sea, approximately 250 kilometers off the northwest coast of Australia. Patches of sheen or weathered oil could have affected at various times an area as large as 90,000 square kilometers. While attempting to plug the leak, a fire broke out on the West Atlas drilling rig. Once more, no one was injured, but the rig was badly damaged.*

Long-term impact

The following quotations are from the Executive Summary of the Commission's report.

> *The Inquiry has concluded that PTTEPAA [the operator] did not observe sensible oilfield practices at the Montara Oilfield. Major shortcomings in the company's procedures were widespread and systemic, directly leading to the Blowout.*
>
> *Well control practices approved by the delegate of the Designated Authority (DA), he Northern Territory Department of Resources (the NT DoR), most likely would have been sufficient to prevent the Blowout if PTTEPAA had adhered to them and to its own Well Construction Standards. However, the NT DoR was not a sufficiently diligent regulator: it should not have approved the Phase 1B Drilling Program for the Montara Oilfield in July 2009 as it did not reflect sensible oilfield practice; it also adopted a minimalist approach to its regulatory responsibilities. The way the regulator (the NT DoR) conducted its responsibilities gave it little chance of discovering PTTEPAA's poor practices. In this case, the regulatory dog did not bark.*

The report was critical not only of the operating company, but also of the regulator who had oversight of this operation. Consequently, one of the report's recommendations was to transfer regulatory authority for installations of this type to the National Offshore Petroleum Safety Authority (NOPSA).

Lessons learned

The causes of this incident are reminiscent of the causes of the Santa Barbara spill: both could have been avoided had the drilling team simply followed standard rules and procedures. The event was not caused by technical limitations or the use of new technology. Related to the failure to follow rules was a failure of the regulatory to fully apply those rules.

The event also highlighted the high public impact that large blowouts can have—a lesson that was to be repeated far more forcefully the following year in the GoM.

Elements of SEMS

Table 2.9 shows the elements of SEMP/SEMS that are of particular relevance to this incident.

Hazards analysis

Management did not carry out a risk analysis. Had they done so, and had they followed up on any findings, the event could have been avoided. For example, management relied on a single barrier (they did not install a second cement plug when mud was removed).

Training

The word "training" is used here in its widest context. The managers and supervisors did not have an adequate plan and were not sufficiently trained to take the proper actions.

Table 2.9 Elements of SEMS

(1)	Safety and Environmental Information
(2)	***Hazards Analysis***
(3)	Operating Procedures
(4)	***Training***
(5)	Pre-Startup Review
(6)	Assurance of Quality and Mechanical Integrity of Equipment
(7)	Safe Work Practices
(8)	Management of Change
(9)	Investigation of Incidents
(10)	Emergency Response and Control
(11)	Audit of Safety and Environmental Management Program Elements
(12)	Records and Documentation

DEEPWATER HORIZON (2010)

The Deepwater Horizon (DWH) event occurred in April 2010 in the GoM. At the time of writing some of the investigation reports (such as that from the Chemical Safety Board) had not been published, so it is important not to draw too many conclusions until those reports and their associated findings are made available. However, it is already clear that this event will have an impact on the offshore oil and gas industry as large as that of Piper Alpha.

In January 2011, Tony Hayward, who was CEO of BP at the time of the DWH event, said that it was a "Black Swan" event. If he meant that the event was "out of the blue" or just bad luck, then he was using the phrase as an excuse. However, the three criteria that Nicholas Taleb listed—the author of the book *Black Swan*—do, in fact, fit the DWH event. This topic is further discussed in Chapter 8.

THE EVENT

DWH was a Mobile Offshore Drilling Unit (MODU) working the Macondo well off the coast of Louisiana. The water depth was approximately one mile. From a technical point of view the background to the incident was quite similar to the Blackbeard situation that has already been described, although the depth of the well was less.

Over a 2-day period the drilling crew faced a series of problems indicating a loss of control of the well. Eventually, there was a blowout (which the blowout preventer failed to stop); oil and gas flowed up the annulus on to the rig where there was an explosion and fire. Eleven of the 126 persons on board died and the drilling rig was destroyed. The oil and gas continued to flow from the badly damaged well head for about 2 months. A record amount of oil was spilled into the ocean.

Long-term impact

The most dramatic long-term consequence was the public relations impact of the ongoing leak taking place a mile below the sea's surface. The daily television pictures dramatized the high-tech nature of the offshore exploration business. But they also showed that the industry did not have the equipment in place to handle such an incident at short notice. Indeed, the equipment needed to bring the spill under control had to be fabricated from scratch.

Many of the organizations involved came out of the event with damaged reputations. Indeed, the regulatory agency that has most to do with offshore safety—the Minerals Management Service (MMS)—renamed itself as the Bureau of Ocean Energy Management, Regulation, and Enforcement (BOEMRE) some three months later. The agency also instituted sweeping management changes.

A more profound lesson learned is that this event—like most other industrial accidents—could easily have been avoided. The drilling crew had received many indications over a 2-day period that things were going awry. If they had been willing to "take a risk for safety" and shut the rig until everything was under control, those men would have lived and the rig would still be floating.

Lessons learned

As already noted, the full impact of this event is yet to be determined. However, based on what has been reported to date the following conclusions can be drawn.

Rules should be followed

The rig was operating in very deep water, but it was not at the "edge of technology." Had the persons on the rig (and their immediate onshore supervisors) simply followed the rules, it is likely that the event would never have occurred.

Not taking a risk for safety

An issue that threads through the entire incident is that the persons on the rig, and their managers onshore, over-rode warning signs and assumed the best when ambiguous results were reported. Rather than "Taking a Risk for Safety" they continued with the drilling program assuming that the situation was under control. So, a partially successful negative pressure test was seen as being a success, not a failure. The persons involved failed to assume the worst.

Elements of SEMS

Given that this incident is still ongoing and under investigation, it is too early to try to analyze in terms of the elements of SEMS. However, some suggestions on those lines are highlighted in Table 2.10.

Table 2.10 Elements of SEMS

(1)	Safety and Environmental Information
(2)	***Hazards Analysis***
(3)	Operating Procedures
(4)	***Training***
(5)	Pre-Startup Review
(6)	Assurance of Quality and Mechanical Integrity of Equipment
(7)	Safe Work Practices
(8)	Management of Change
(9)	Investigation of Incidents
(10)	***Emergency Response and Control***
(11)	Audit of Safety and Environmental Management Program Elements
(12)	Records and Documentation

FUKUSHIMA DAI-ICHI (2011)

The causes and consequences of the nuclear accident at Fukushima Dai-ichi were still being established at the time of writing. Yet, even though the details as to what occurred are not yet fully known and the final consequences are to be determined, it is already clear that the event is of such severity that it will impact the energy industry world-wide. As with the Three Mile Island event, even though this incident is to do with nuclear power, it is likely to have an impact on the offshore oil and gas industry.

The event

In March 2011, a once-in-a-thousand-years offshore earthquake struck northern Japan. The earthquake was followed by a massive tsunami. The two events caused immense loss of life and damage to the nation's infrastructure. The earthquake and tsunami also caused serious damage to at least two of the nuclear reactors at the Fukushima Dai-ichi complex and one of the pools used for storing spent fuel rods.

Most of the damage to the nuclear power plant came from the 14 m tsunami that overwhelmed the totally inadequate 5.5 m seawall. The seawater knocked out the safety systems designed to keep the reactors cool. Consequently, the cores of two of the reactors overheated, leading to what was probably a partial meltdown. This in turn led to follow-on problems, such as the generation of hydrogen gas that then exploded (there is some speculation that at least one of the explosions was nuclear). A considerable amount of radioactive material leaked to the ground, the sea, and the air—and those leaks may be ongoing.

Long-term impact

It is already clear, even before the event has been brought to a closure, that the Fukushima event will have profound consequences, some of which are listed below.

- Many people living and working in the general area of the damaged nuclear power plants may suffer illnesses induced by radiation exposure. This applies particularly to the workers involved in the effort to stabilize and control the cores. The Japanese nuclear power authorities had set a target of one radiation-induced fatality in a million years just eight years prior to the accident. Yet already it appears as if radiation from the stricken units will lead to sickness and possible early death of a considerable number of residents and workers.
- A substantial area around the plant may have to be evacuated for many years, and sales of agricultural products have been restructured.
- Contamination of the ocean has led to a restriction in commercial fishing.
- The cost of making the site safe for generations to come will be enormous.
- The Fukushima nuclear power plants will never generate electricity commercially again. This means that the Japanese economy will suffer shortages of electrical power for many years to come, even if oil or gas-based replacement power plants can be quickly built and commissioned.
- World-wide, public confidence in nuclear power has fallen, such that, in Germany, a decision has been made to phase out all of the nation's 17 nuclear facilities. Partly, this loss of confidence has occurred due to the increasing awareness of some of the long-term problems associated with nuclear power—particularly what to do with the disposal of thousands of tonnes of radioactive waste.
- There will be much greater public scrutiny of all nuclear activities, and the costs of building and commissioning new plants will increase substantially. Indeed, the future of the very industry may be threatened.

None of the above issues directly affect the offshore oil and gas industry, but they do impact the energy business in general. Some of the potential impacts are listed below.

- If the use of nuclear power declines, there will be a greater demand for other types of fuel, particularly natural gas, needed to generate electricity.
- Public confidence in advanced technology, which had already been dented as a result of the DWH incident, will go down even further.
- The Fukushima event heightens awareness of "black swan" or long-tail events, as discussed in Chapter 8.

Lessons learned

It is too early to develop long-term lessons learned, particularly as they apply to the oil and gas industry, except to say that technologically-challenging ventures, whether they are nuclear power plants or deepwater drilling platforms, will be subjected to much

Table 2.11 Elements of SEMS

(1)	Safety and Environmental Information
(2)	***Hazards Analysis***
(3)	Operating Procedures
(4)	Training
(5)	Pre-Startup Review
(6)	***Assurance of Quality and Mechanical Integrity of Equipment***
(7)	Safe Work Practices
(8)	Management of Change
(9)	Investigation of Incidents
(10)	***Emergency Response and Control***
(11)	Audit of Safety and Environmental Management Program Elements
(12)	Records and Documentation

more scrutiny, and managers in those industries will have to conduct many more safety analyses and formal safety assessments.

Another lesson learned is to do with public relations which, in the case of Fukushima Dai-ichi, were not good.

Elements of SEMS

Given that this incident is still ongoing and under investigation, it is too early to try to analyze in terms of the elements of SEMS. However, some suggestions on those lines are highlighted in Table 2.11.

Hazards analysis

A thorough hazards analysis would have quickly identified the dangers associated with tsunamis. (It is possible that the tsunami problem was identified, but then not followed up on, in which case the hazard analysis tracking system would have helped ensure that this hazard would be properly addressed.)

Emergency response and control

The response to the incident was unimpressive. It appears as if senior management was reluctant in the days following the event to add seawater to the reactors because doing so would have put those reactors out of service forever. (In the end, they did use seawater, but only after a considerable amount of damage had been done.)

REPORT OF THE DWH PRESIDENT'S COMMISSION

In the wake of the DWH tragedy, US President Obama set up a commission to look into its causes, and to make recommendations as to how such events could be avoided in the future.

The committee issued its report in January, 2011, although critical information—particularly the conditions of the Blowout Preventer—was not known at the time of publication.

The report's scope is broad, going beyond just technical issues. For example, the report provides extensive discussion on the economy of the GoM, and the impact of the spill on the people who work there. The discussion here focuses just on those items in the report that affect technical and PSM issues (National Commission, 2011).

The report is voluminous—some 340 pages. It is divided into the following 10 chapters, the titles of which, as can be seen, are quotations from those involved in the event or its aftermath:

(1) Chapter 1 "Everyone involved with the job was completely satisfied"

(2) Chapter 2 "Each oil well has its own personality"

(3) Chapter 3 "It was like pulling teeth"

(4) Chapter 4 "But, who cares, it's done, end of story…

(5) Chapter 5 "You're in it now, up to your neck!"

(6) Chapter 6 "The worst environmental disaster America has ever faced"

(7) Chapter 7 "People have plan fatigue… they've been planned to death"

(8) Chapter 8 "Safety is not proprietary"

(9) Chapter 9 "Develop options…" (this chapter contains the commission's recommendations)

(10) Chapter 10 "American Energy Policy and Future of Offshore Drilling"

There are six appendices.

The report starts with the statement, "Complex Systems Almost Always Fail in Complex Ways." In other words, there was no single cause for the catastrophe. Another sentence in the report's Foreword amplifies the above sentence in human terms, "… the business culture succumbed to a false sense of security." It goes on to say, "… that complacency affected government as well as industry."

The following quotations best summarize the "Overarching Management Failures by Industry" that, in the judgment of the report's writers, led to the event.

> *The blowout was not the product of a series of aberrational decisions made by a rogue industry or government officials that could not have been anticipated… Rather, the root causes are systemic and, absent significant reform in both industry practices and government policies, might well recur.*
>
> *The most significant failure at Macondo—and the clear root cause of the blowout—was a failure of industry management… Better management of decision making processes within BP and other companies, better communication within and between BP and its contractors, and effective training of key engineering and rig personnel would have prevented the Macondo incident.*
>
> *[…] operators must have effective systems in place for integrating the various corporate cultures, internal procedures, and decision making protocols of the many different contractors.*

Root causes

The report has a section entitled, "The Root Causes: Failures in Industry and Government." A brief summary of these root causes is provided below. The detailed recommendations are discussed in the next section.

Management of change

BP's management process did not adequately identify or address risks created by late changes to well design and procedures.

Many process safety professionals feel that the topic of MOC is the most difficult to implement.

> *[...] changes to drilling procedures in the weeks and days before implementation are clearly not subject to any[...] MOC process.*

The report particularly notes that temporary abandonment procedures did not go through "any formal review at all." This comment highlights one of the biggest problems to do with MOC, which is how to handle temporary changes.

> *From a safety and operational point of view, whether or not a change is permanent or temporary is merely a semantic matter—the system itself does not know or care that a change is intended to be temporary. Therefore, the fact that a proposed change is defined as being "temporary" does not mean that it can be handled less rigorously than a change that is intended to be permanent. Yet, because of the short duration of temporary changes, the personnel implementing them may be tempted to take short cuts, particularly if going through the MOC process takes longer than actually making the change itself. There is a temptation to take an attitude of "let's just get on with it—why bother spending hours writing and reviewing a procedure for an operation that will only take a few minutes to carry out?"*

Intervention

One of the tenets of behavioral-based safety programs is that of "intervention." If someone observes an unsafe condition, it is his or her responsibility to step up and point out the problem. (On the day that this was written the author had one of his colleagues intervene. There had just been an office move and she pointed out that the chaotic state of his furniture would make escape troublesome in the event of an emergency).

With regard to the DWH it was apparent to many that there were problems with the integrity of the cement being used, but no one "intervened." No one showed the leadership necessary to take actions that might have ended up with them in trouble.

Communication

There was a lack of communication at all levels and between all parties regarding the anomalous data that was being observed. It appears as if the lines of command were not properly defined and/or being followed. (This is contrast to the Blackbeard incident discussed where the difficulties that the drill crew was experiencing were communicated all the way up the command chain, stopping only at the company's CEO.)

Lessons learned

Transocean, the owner of the drilling rig that was leased to BP, had experienced "an eerily similar near-miss on one of its rigs in the North Sea four months prior to the Macondo blowout." In that incident the mud was spewing on to the rig floor, but the crew was able to shut in the well before it became a full blowout. Transocean created an internal presentation regarding this incident but the DWH drill team never saw it.

Balancing risk and profitability

Figure 4.10 in the Commission's report identifies nine decisions that were made at Macondo that increased the level of risk. In all cases a less risky alternative was available. However, in at least seven of the cases, the more risky decision meant that time could be saved.

The report faults management at Macondo for not having a formal system to ensure that lower cost or time-saving decisions could be made without increasing risk to safety and the environment. This problem relates to the MOC and intervention issues already alluded to.

Regulatory failures

The report notes that many of the critical decisions and mistakes that were made on the DWH rig were not covered by regulations. It attributes the weakness in regulatory oversight to a lack of government commitment stretching over many years, and to the lack of expertise and training within the MMS.

Recommendations

The report of the President's Commission regarding the DWH event contains a total of 29 recommendations. They are divided into the following seven categories:

A. Improving the Safety of Offshore Operations

B. Safeguarding the Environment

C. Strengthening Oil Spill Response, Planning, and Capacity

D. Advancing Well-Containment Capabilities

E. Overcoming the Impacts of the DWH Spill and Restoring the Gulf

F. Ensuring Financial Responsibility

G. Promoting Congressional Engagement to Ensure Responsible Offshore Drilling

The recommendations for Sections A to D are listed below. For some of them comments that are pertinent to the substance of this book are provided.

NOTE: In the interests of space, some of the recommendations and associated text have been condensed—it goes without saying that anyone who wishes to know more about a particular recommendation should refer to the original, complete text of the report.

A1: Risk management

The Department of the Interior should supplement the risk-management program with prescriptive safety and pollution-preventing standards.

As discussed in Chapter 1, the management of risk offshore involves an ongoing debate between prescriptive and nonprescriptive approaches to safety management. The trend for decades, as evidenced in the Cullen report, has been toward non-prescriptive standards. The report suggests that it still remains important to understand that prescriptive rules play an important role.

A2: "Safety case" approach

The Department of the Interior should develop a proactive, risk-based performance approach specific to individual facilities, operations and environments, similar to the "safety case" approach in the North Sea.

Chapter 6 of this book describes the safety case approach to managing risk. One of the reasons that this approach was not followed in the GoM following the Piper Alpha event is that there are thousands of platforms in the Gulf, many of them small, unmanned and very similar to one another. It would not be feasible to develop a safety case for each of these.

Prior to the introduction of the final SEMS rule there has been some speculation that the MMS/BOEMRE would require that safety cases be prepared for deepwater platforms. Many companies are developing risk-based analyses of these platforms anyway because their loss would be so costly. Therefore, a safety case requirement for those platforms may not be as demanding as it may sound.

This recommendation also calls for a risk-based approach to be used. Such is not the case at present with regard to either SEMP or SEMS.

A3: International forum

Working with the International Regulators' Forum and other organizations[...] identify those drilling, production, and emergency-response standards that best protect offshore workers and the environment, and initiate new standards and revisions to fill the gaps and correct deficiencies. These standards should be applied[...] globally wherever the international industry operates. Standards should be updated at least every five years as under the formal review process of the International Organization for Standardization (ISO).

To some extent standards from the API, such as Recommended Practice 14C, are already filling this role.

Supplemental to the above recommendation, the following actions are suggested:

- An engineering consultant should determine what constitutes the highest level of current standards, and set up a certification process.
- Safety cases should be developed as part of exploration and production plans.
- SEMS programs should include regular third-party audits.
- Ensure that the government has sufficient expertise in the area of drilling safety. (Detail is provided on how this recommendation could be executed.)
- Develop more detailed requirements for incident reporting, including near-misses.

A4: Independent agency

Congress and the Department of the Interior should create an independent agency within the Department of the Interior with enforcement authority to oversee all aspects of offshore drilling safety (operational and occupational), as well as the structural and operational integrity of all offshore energy production facilities, including both oil and gas production and renewable energy production.

The report is indicating that the transition from MMS to BOEMRE was a step in the right direction, but that further improvements to the enforcement of rules and regulations are required.

A5: Regulatory funding

Congress and the Department of the Interior should provide a mechanism[...] for the payment of regulatory fees, for adequate, stable, and secure funding to the key regulatory agencies—Interior, Coast Guard, and NOAA.

This recommendation addresses not only the need to keep safety oversight independent of other regulatory roles, but it also discusses the need for a systemic reorganization that would address many of the problems that arise at regulatory interfaces.

This section also discusses a Proposed Reorganization of the former MMS into a new organization. The report recommends that the head of the new agency should have a "relevant engineering or technical background."

B1: Strengthen NEPA

The Council on Environmental Quality and the Department of the Interior should revise and strengthen the NEPA (National Environmental Protection Act) policies, practices, and procedures to improve the level of environmental analysis.

B2: Interagency activities

The Department of the Interior should reduce risk to the environment from OCS oil and gas activities by strengthening science and interagency consultations.

C1: Oil spill planning

The Department of the Interior should create a rigorous, transparent, and meaningful oil spill risk analysis and planning process for the development and implement of better oil spill response.

Industry has had two responses to the above requirement in the aftermath of the DWH incident. They are the *Helix Fast Response System* and the *Marine Well-Containment System (MWCC)*. The BOEMRE required that companies use one of these two systems before drilling permits could be issued following the DWH catastrophe. Both companies reported that their systems were ready in February 2011, although the MWCC System will undergo much more development.

The recommendation requires that owners/operators develop realistic plans for responding to worst-case situations. (One of the more embarrassing features of the DWH incident was the cookie-cutter nature of the oil spill plans of various companies operating in the GoM. These plans included references to seals and walruses—animals that have not been present in the area for many thousands of years.)

This recommendation also stresses the need for the regulatory agencies to understand the oil spill response programs so that such plans can be evaluated properly.

C2: District plans

EPA and the Coast Guard should establish distinct plans and procedures for responding to a "Spill of National Significance."

C3: State and local involvement

EPA and the Coast Guard should bolster state and local involvement in oil spill contingency planning and training and create a mechanism for local involvement in spill planning and response similar to the Regional Citizens' Advisory Councils mandated by the Oil Pollution Act of 1990.

C4: Increased research

Congress should provide mandatory funding for oil spill response research and development and provide incentives for private-sector research and development.

C5: Dispersants

EPA should update and periodically review its dispersant testing protocols for product listing or pre-approval, and modify the pre-approval process to include temporal duration, spatial reach, and volume of the spill.

C6: Offshore barrier berms

The Coast Guard should issue guidance to establish that offshore barrier berms and similar dredged barriers generally will not be authorized as an oil spill response measure in the National Contingency Plan or any Area Contingency Plan.

D1: Oversee source-control efforts

The National Response Team should develop and maintain expertise within the Federal government to oversee source-control efforts.

D2: Source-control plans

The Department of the Interior should require offshore operators to provide detailed plans for source control as part of their oil spill response plans and applications for permits to drill.

D3: Accurate estimates

The National Response Team should develop and maintain expertise within the federal government to obtain accurate estimates of flow rate or spill volume early in a source-control effort.

D4: Proposed well design

The Department of the Interior should require offshore operators seeking its approval of proposed well design to demonstrate that:

- Well components, including blowout preventer stacks, are equipped with sensors.
- Wells are designed to mitigate risks towel integrity during post-blowout containment efforts.

The importance of the last point can be seen with the Santa Barbara spill of 1969 (p. 49) in which pressure from the well led to failure of seabed located miles from the original leak site.

INPO

Although not written up as a recommendation, the Commission's report discusses the "Nuclear Model" at some length. This discussion seems to be leading, at the time of writing, to the creation of a Center for Offshore Safety under the aegis of the API. The role of this new organization, and how it compares to what occurred in the nuclear power industry following the Three Mile Island disaster, is discussed in the next chapter.

NEED FOR NEW STANDARDS

Since one of the focuses of this book is on the development and application of regulations and standards—specifically SEMP and SEMS—it is useful to examine the events listed above to see how rules and regulations affected their occurrence. Some thoughts on this topic are provided in Table 2.12.

The conclusion that can be drawn from Table 2.12 is that failure to follow rules and regulations was a major factor in four of the listed events (with the obvious exception of Blackbeard). As the President's Commission report on DWH pointed out, it is imperative that the rules are modified and upgraded to meet the challenges posed by new technology. However, the reality is that many events could have been avoided had the facilities involved simply followed the old, established regulations.

Table 2.12 Regulatory Impact and Role

Event	Discussion
Santa Barbara	The federal regulations were sufficient so this event could have been prevented. One of the root causes of the event was failure to follow those regulations, and a willingness on the part of the regulators to allow the operator to take a variance from the standards.
Piper Alpha	This event cannot be attributed to failure to follow regulations. Indeed, the Cullen report led to major developments in technical safety and formal safety analyses and of regulations built around the safety case concept.

Table 2.12 Regulatory Impact and Role—*Cont'd*

Event	Discussion
Valdez	Had the ship's instrumentation been working as called for, it is likely that this event would not have occurred. This failure may not have been a regulatory matter *per se*, but it showed an overall disregard for following the rules.
Blackbeard	Management's decision to stop this operation did not involve regulations or the regulators.
Montara	Failure to implement standard regulatory practices led to the occurrence of this event. It is similar to Santa Barbara in that following the rules would have been enough to prevent the blowout.
Deepwater Horizon	Although final reports on this event are not yet published, it appears as if the drilling crew bypassed standard operating practices. The various investigations that are under way will determine whether regulations and rules were bypassed.
Fukushima Dai-ichi	There have been extensive discussions in the press to do with the close relationship between the Japanese nuclear power industry and the regulators. (The phrase "nuclear village" is sometimes used.) It appears as if there is a need not so much for new regulations, but for an environment in which existing regulations are enforced openly and independently.

REFERENCES

Montara Commission of Inquiry, June 2010. Commonwealth of Australia.

National Commission on the BP Deepwater Horizon Oil Spill and Offshore Drilling. January 2011.

CHAPTER

Safety and environmental management programs

3

CHAPTER CONTENTS

INTRODUCTION

In Chapter 1, it was noted that, following the Piper Alpha catastrophe of 1988, two approaches to the development of Safety Management Systems (SMSs) were followed. The first (shown in the top line of Figure 1.5) was the use of Safety and Environmental Management Program (SEMP; and later SEMS) standards, mostly in the USA. The second approach, used initially in the United Kingdom, and later elsewhere, was that of Safety cases. This chapter discusses the first of these approaches: SEMP and the use of industry standards. Safety cases are described in Chapter 6.

At the time of the Piper Alpha event, the American offshore oil and gas industry, which is concentrated in the Gulf of Mexico (GoM), consisted of hundreds of small platforms, most of them in shallow water, and all having a generally similar design. Therefore, the industry decided to develop a standard that could be applied to all of these smaller facilities, and that would not require each of them to develop their own unique analysis and safety program. This decision led to the development of the American Petroleum Institute's (API) Recommended Practice 75—*Development of a Safety and Environmental Management Program for Offshore Operations and Facilities*. This standard, referred to as RP 75 throughout the remainder of this book, calls on API member companies to develop and implement a Safety and Environmental Management Program (SEMP). (In practice, the terms "RP 75" and "SEMP" can be used interchangeably.)

Ian Sutton: Offshore Safety Management. DOI: 10.1016/B978-1-4377-3524-6.00003-4
Copyright © 2012 Ian Sutton. Published by Elsevier Inc. All rights reserved.

API STANDARDS

The American Petroleum Institute (API) represents about 400 companies in the energy business. Activities of these companies include offshore oil and gas production, oil refining, and the distribution of finished hydrocarbon products. Although API is best known for its representation of industry interests to the public and to the government, it also has a technical arm that develops technical standards, many of which are used around the world. Some of these standards are known as Recommended Practices (RPs). Because these practices are only recommended, companies are not generally legally obliged to follow their requirements of the RPs. (However, some of them have been incorporated into various USA rules and regulations.) The following quotation is from the API website (www.api.org):

> *The development of consensus standards is one of API's oldest and most successful programs. Beginning with its first standards in 1924, API now maintains some 500 standards covering all segments of the oil and gas industry. Today, the API standards program has gone global, through active involvement with the International Organization for Standardization (ISO) and other international bodies*

CENTER FOR OFFSHORE SAFETY

In the previous chapter, it was noted that the report to the president concerning the Deepwater Horizon catastrophe discussed the creation of an industry-sponsored agency analogous to the nuclear industry's independent Institute of Nuclear Power Operations (INPO) that was created in the wake of the Three Mile Island event. The President's Commission to do with *that* event said the following.

> *[T]he nuclear industry must dramatically change its attitudes toward safety and regulations. The Commission has recommended that the new regulatory agency prescribe strict standards. At the same time the industry must also set and police its own standards of excellence to ensure the effective management and safe operation of nuclear power plants.*

There are three key elements to the above quotation as applied to the offshore oil and gas industry.

The first is to do with industry's attitude to regulations. In the next chapter it is shown how, at least up until the time of the Deepwater Horizon incident, many representatives of the industry were quite resistant to the need for increased regulations—specifically the need for SEMS. The Commission's report suggests that a change is needed. After all, the oil industry does not own the Outer Continental Shelf. The federal government has the authority and responsibility to require industry to take whatever actions are needed to achieve safe operations.

However, the quotation about the nuclear power industry recognizes that industry should set its own standards. The nuclear power industry did this through the creation of INPO. A similar organization in the offshore oil and gas industry would set standards, require that all facilities meet those standards, measure performance, and compare facilities and companies with one another. Such actions would help ensure that the whole industry is not sullied by the under-performance of a just a few facilities. Moreover, because the offshore industry is so technically sophisticated, it is unlikely that the government regulators will ever possess all of the expertise needed to fully control the industry. Therefore, self-regulation in the manner in which it is carried out by INPO is a sensible solution to this difficulty.

A key role of INPO is that it carries out thorough, but nonadversarial, inspections of nuclear power plants. The inspectors are often employees of other power plants on loan to INPO. This practice encourages valuable cross-fertilization of ideas. The results of the inspections are confidential, and shared only with the managers at the affected sites. Although some mechanism is needed for protecting confidential information, such an approach could be a valuable way of improving offshore safety.

It must be recognized, however, that there are fundamental differences between the two industries. Hence, the model that works for the nuclear power industry may not be entirely appropriate for offshore operations. Some of these differences include the following:

- Whereas the nuclear power industry has a small number of plants operating in much the same way year after year, the offshore industry is much more dynamic. Rigs and platforms are constantly moving to new locations, and the operating conditions at each well are unique.
- The oil and gas industry is much more fragmented than is the nuclear power industry. There are 104 nuclear power plants in the USA. On the other hand, there are thousands of companies operating offshore in one capacity or another.
- Oil and gas companies are in competition with one another in a manner that is not true for public utilities. The creation of an INPO-type organization, particularly with regard to exploration and drilling, would meet strong resistance because so much information is proprietary and/or a trade secret.

Given this background, the American Petroleum Institute (API) created the Center for Offshore Safety (COS) following *Deepwater Horizon*. The center will supplement formal regulations; and will have a relationship with BOEMRE and the other agencies analogous to that between INPO and the Nuclear Regulatory Commission.

In March 2011, the API stated the following regarding the proposed Center:

> *After extensive review and development, the oil and natural gas industry has approved the creation of the Center for Offshore Safety, which will promote the highest level of safety for offshore operations, through an effective program that addresses management practices, communication and teamwork, and which relies on independent, third-party auditing and verification. The board directed API to further develop the operational framework and timeline for the center—working with other industry stakeholders—to enhance industry safety and environmental performance.*
>
> *Today's actions follow API's August 2010 testimony and its October 2010 letter to the Presidential Oil Spill Commission, regarding the industry's review of five other safety programs, including: the U.K.'s Step Change in Safety, the chemical industry's Responsible Care® program, the Institute for Nuclear Power*

Operations, the OSHA Voluntary Protection Program, and the Safety Case Regime for international operators.

The Center for Offshore Safety will draw on the lessons learned from successful, existing safety programs, applying the best elements of these programs to accommodate the unique challenges of offshore oil and natural gas operations, and will rely on independent auditing and review by third-party certifiers, according to API.

While many of the Center's structure and operating details have yet to be resolved, the program will be run by the separately-funded standards and certification arm of the American Petroleum Institute, which receive regular audits and accreditation by the American National Standards Institute and the ANSI-ASQ National Accreditation Board.

The Center will be based in Houston and will be open to all companies that operate in deepwater exploration and production. The focus of the center will be based on API's Recommended Practice 75, covering safety and environmental management systems, which was recently incorporated into federal regulations by BOEMRE.

The COS has set the following goals for itself:

- Provide a platform for collaboration on safety issues
- Have full industry participation
- Provide independent audit and certification services
- Provide tools for continuous improvement
- Develop techniques for operational excellence
- Provide expert assistance

Because API represents industry, it was important that the center was seen as being independent, which is why it will be audited and accredited by other organizations. It is likely that BOEMRE will not have a person on the board. It is planned for the center to be formally launched in the third quarter of 2011, with offices located in downtown Houston.

A challenge for the center is that many other organizations—including individual companies and BOEMRE—will be conducting audits at the same time. Hence, the demand for qualified independent third party auditors (I3Ps) is likely to exceed the supply.

OFFSHORE OPERATORS COMMITTEE

The Offshore Operators Committee (OOC) represents oil and gas companies working in the Gulf of Mexico and Atlantic Ocean (but not California). The organization says the following about itself:

The Offshore Operators Committee (OOC) is a non profit organization comprised of any person, firm or corporation owning offshore leases and any person, firm or corporation engaged in offshore activity as a drilling contractor, service company, supplier or other capacity that desires to participate in the work of OOC or the Offshore Operators Committee.

The Committee's activities are focused on providing its member operators with information and technical support that will assist them in conducting their offshore activities in a manner that will promote sound safety and environmental operational practice.

The OOC was active in challenging the need for the first version of SEMS in 2009. During 2011, the organization has set up various committees and subcommittees to develop means of working with SEMS.

FUNDAMENTALS OF SAFETY MANAGEMENT SYSTEMS

All SMSs share fundamental features, regardless of technology or location. Some of these features are listed below, and then discussed in the following sections.

- Safe limits
- Nonprescriptive
- Risk based
- Involvement and thoroughness
- Holistic

Safe limits

The safe limits for each process variable must be defined quantitatively. For example, the safe temperature range for operating a compressor may be 125–150 °C. If the actual temperature deviates outside of that range, then that operation is—by definition—out of control and potentially unsafe; action *must* be taken to bring the temperature back into the correct range. The fact that the process has deviated outside the safe range does not mean that an emergency situation exists—there may be plenty of time to take action. But doing nothing is not an option.

Once the safe range has been defined, management must determine how to operate their facility so that it stays within that range. In the case of the compressor temperature example, instrument set points must be adjusted and operators trained to achieve the 125–150 °C range. All the people involved in running or maintaining the unit must know how to identify an out-of-control situation, what its consequences might be, and how they should respond to it. If it is management's intention to operate outside the prescribed range, then the Management of Change program should be implemented in order to ensure that the new conditions are safe, new limits have been set, or new safeguards have been installed.

When a facility is new, the safe limits are defined by its designers. As operating experience is accumulated, new safe limit values will be implemented—often through use of the hazards analysis and management of change processes.

Some safe limits may have no meaningful value. For example, if a pressure vessel is designed for full vacuum operation, then that vessel has no safe lower limit for pressure.

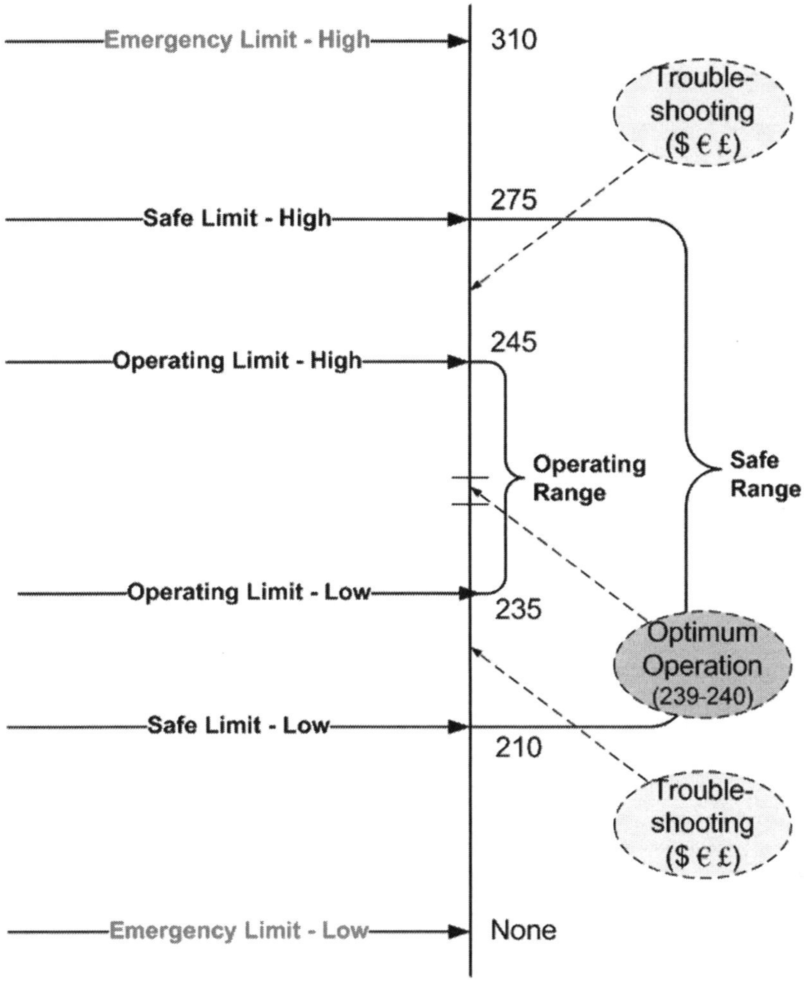

FIGURE 3.1

Operating, Safe, and Emergency Limits.

The concept of safe limits can be extended to include operating and emergency limits, as illustrated in Figure 3.1, which shows values for process variables such as pressure, temperature, level, or flow rate.

The innermost range of Figure 3.1 shows the *optimum* value for this particular parameter. In this case, it is 239–240. This optimum point may change as target conditions concerning production rates, yields, or product quality change.

The *operating range* represents the upper and lower limits for that variable's normal value. Supervision is free to move the variable to any point within that range in order to achieve production and quality goals. In Figure 3.1, the operating range is 235–245.

If operating conditions are allowed to move outside the operating limits, but within the safe limits, then the facility is said to be in "trouble," i.e. there are no safety issues to worry about, but the system is operating inefficiently. Troubleshooting efforts to bring the value back into the operating range will save money. Indeed, much of management's attention will be directed toward troubleshooting because addressing difficulties in this area will often lead to a significant improvement in profitability for relatively little expenditure. Examples of "trouble" include:

- Excessive energy consumption
- Product quality problems
- Unusually high use of spare parts
- Low production rates

The operating limit values are often quite fuzzy. As the system moves away from optimum operation, it will start to exhibit symptoms of unusual operation which will eventually lead into the troubleshooting range.

The next range is defined by the *safe limit* values. In Figure 3.1 where the parameter is allowed to exceed 275 or go below 210, then the system is in an unsafe condition and action must be taken to bring that value back into the safe range.

The final set of values is the *emergency limits*. If the process parameter goes beyond one of these limits, then an emergency situation has been created. Immediate action is required; generally, the safety instrumentation and safety equipment (such as pressure relief valves) will be activated. In Figure 3.1, the upper emergency limit is 310; there is no lower emergency limit in this example.

The relationship between operating, safety, and emergency limits is shown in Table 3.1. The fourth column in Table 3.1 describes the actions taken if the emergency gets out of hand, and emergency response teams have to be mobilized.

Nonprescriptive

SMSs are largely nonprescriptive; i.e. the regulations and standards in this field generally provide very little specific detail as to what has to be done. Basically they say, "Do whatever it takes *on your facility* not to have accidents." It is up to the managers, technical experts, and the operations/maintenance personnel to determine how this should be done. This lack of detail explains why the technical section of the SEMS rule is so short. The regulation merely requires that programs be in place, that they be adhered to, and that they work.

Although each facility is unique, many operations, such as starting a pump or training a contract worker, are really quite similar from site to site and from company to company. Therefore it is possible to develop high quality, generic systems that can be used in a wide variety of situations. Doing so saves time and money, and improves quality. It is this philosophy that lies behind the creation of RP 75 and the many prescriptive standards that it calls on.

Performance based

Nonprescriptive management programs have to be performance based because the only measure of success is success. Hence, the only true measure of success of the program is not to have incidents. But, from a theoretical point of view, such a goal is

Table 3.1 Types of Nonstandard or Abnormal Situation

Operational Deviation	Safety Deviation	Emergency Operation	Emergency Response
Limit Values			
The operation stays within the safe limits.	Some operating parameters move outside their safe limits, but not at the emergency level. Time is not of the essence.	The emergency limits are exceeded; emergency operations and/or automated instrument response are required.	The emergency has spread to other units.
Severity of Consequences			
The consequences of the problem are primarily economic, although failure to address the situation may lead to a safety problem eventually.	The consequences resulting from the deviation are that worker safety is jeopardized and/or a major environmental problem may result.	The deviation is very serious. There is an immediate danger of a fatality or of a major environmental release.	The situation has deteriorated so that an entire facility is threatened, not just one operating unit. The public may also be affected.
Response Time			
Usually, there is time to review what needs to be done.	Action must be taken because safety standards have been violated. However, there may be plenty of time to evaluate what needs to be done.	Speed is essential.	Speed is essential.
Operating Procedures Requirements			
A Troubleshooting Guide is needed. The instructions can be quite lengthy, discursive, and complex, if necessary.	The instructions can be reasonably detailed, and they can offer options. However, they must be unambiguous.	The instructions must be short in number, simple, and easy to execute.	The instructions will provide guidance to a trained emergency response team.
Different points of view can be presented because there may be different causes that generate the same symptoms and because more than one solution may be viable.The instructions take the form of guidance or suggestions; there is no absolutely correct or incorrect way of addressing the situation.	The instructions must be followed as written. However, there is room for interpretation and judgment.	Absolutely no ambiguity is permitted.	Absolutely no ambiguity is permitted.

impossible to achieve. No matter how well run a facility may be, accidents will occur; risk can never be zero. For this reason, some risk management professionals chose not to use the terms "compliance" on the grounds that true compliance can never be realized. Because risk can never be zero, there are always ways of improving safety and operability.

Holistic

The elements of an SMS have strong interactions with one another—it is not possible to meet the requirements of one of the elements without considering its effect on the others.

The interconnectedness of the elements can be illustrated by considering the development of an Emergency Response Plan, in which the following sequence of actions—involving seven of the elements of SEMS (Table 3.2) —may occur.

- The writing of the *Emergency Response Plan* (element 10) requires a knowledge of which hazards have to be addressed.
- Consequently, a *Hazards Analysis* (element 3) is required to identify the hazards.
- In order to be able to carry out the hazards analysis, information from sources such as P&IDs and MSDS is needed. Much of this information is included in the *Safety and Environmental Information* (element 1).
- Once the Emergency Response Plan has been developed, it will be necessary to *Train* everyone in its use (element 7).
- The Emergency Response Plan has to be *Audited* on a regular basis (element 12).

Table 3.2 Elements of SEMP

(1)	General
(2)	Safety and environmental information
(3)	Hazards analysis
(4)	Management of change
(5)	Operating procedures
(6)	Safe work practices
(7)	Training
(8)	Assurance of quality and mechanical integrity of critical equipment
(9)	Prestart-up review
(10)	Emergency response and control
(11)	Investigation of incidents
(12)	Audit of SEMP elements
(13)	Records and documentation

- During the training process, the trainees will come up with ideas that will improve the quality of the emergency response plan. This is *Workforce Involvement* (General).
- After going through the *Management of Change* step (element 4), these ideas can be used to upgrade the emergency manual.

When considered in isolation, many of the elements appear to be the "most important." For example, *Workforce Involvement* is the "most important" element because, if the employees do not participate, the process safety program will not function properly. But *Management of Change* could be considered the "most important" element because the root cause of all incidents is uncontrolled change. On the other hand, all of the elements require a solid base of up-to-date, comprehensive information. Therefore, *Knowledge Management* is the "most important" element. But then it could be argued that *Incident Investigation and Root Cause Analysis* is what really matters because incidents reveal what is really going on in the organization. The real point, of course, is that they are all important and necessary, and that they all rely on one another to be effective.

Involvement and thoroughness

An SMS is not something that is created and then handed down by management to their employees and contract workers; it is a program that involves everyone: designers, operators, maintenance technicians, managers, and senior executives. The keyword is *involvement*—which incorporates much more than just *communication*. All managers, employees, and contract workers are responsible for the successful implementation of the program. Management, who must provide determined and committed leadership, must organize and lead the initial effort, but the employees must be fully involved in its implementation and improvement because they are the ones who know the most about how a process really operates, and they are the ones who have to implement recommendations and changes. Specialist groups such as staff organizations and consultants can provide help in specific areas, but process safety is fundamentally a line responsibility.

The implementation of an SMS also requires thoroughness. For example, a company may have a good training program, but one person may have missed part of it because he or she was on vacation. Management will have to make sure that this person is trained and that his or her personnel files are updated appropriately.

Both involvement and thoroughness require that those in charge show not only their management skills, but also their leadership skills.

RECOMMENDED PRACTICE 75

Recommended Practice 75 (API, 2004) was first published in the year 1991. It is now in its third edition. It calls on offshore operators to develop a SEMP. RP 75 is written to meet the following criteria:

- It is performance based, with progress generally being measured on an annual basis.

- A system for establishing continuous improvement should be in place.
- Effective communication is required.

Scope

RP 75 is broad in scope. It addresses the identification and management of safety hazards and environmental impacts in the following areas:

- Design
- Construction
- Start-up
- Operation
- Inspection
- Maintenance

It also covers new, existing, or modified drilling and production facilities.

Management

The standard requires that a company put in place a SEMP that meets the following principles:

(1) A written program is required.

(2) Leadership, accountability, and resources must be provided.

(3) Specific representatives need to be appointed to establish, implement, and maintain the SEMP.

(4) There are also representatives who report to management on the performance of the SEMP.

(5) The SEMP should be reviewed regularly.

(6) A written description of the organization and lines of responsibility is required.

(7) The expertise of personnel should be utilized in areas such as the identification of hazards, the development of safe work practices, the development of training programs, and the investigation of incidents.

(8) The owner, operator, and contractor management have their own responsibility for assuring safe operations.

(9) Industry codes should be used in the design, construction, maintenance, and operation of the facility.

(10) The management of safety is an integral part of the design and operation of the facility.

(11) All persons involved in SEMP must be properly trained.

(12) The SEMP should be regularly audited.

ELEMENTS OF SEMP

At the heart of SEMP are 13 management and technical elements. They are listed in Table 3.2. The standard also has five appendices.

Appendix A—Contractor Selection Criteria

Appendix B—Industry Codes, Practices, and Standards

Appendix C—References

Appendix D—Definitions

Appendix E—Performance Measures

The "General" section and the five appendices are discussed in this chapter. A description of the 12 technical elements (2–13) is provided in the next chapter relating to SEMS.

GENERAL

The first section of RP 75, General, discusses general principles of offshore safety management and provides guidance on the scope of the standard and the manner in which it is to be applied. Some of the key provisions of this section are as follows.

Purpose and objective

The scope of RP 75 covers the identification and management of safety hazards and environmental impacts in the design, construction, start-up, operation, inspection, and maintenance of new, existing, and modified drilling and production facilities. Both drilling and production are included.

RP 75 states, "The objective of this recommended practice is to form the basis for a Safety and Environmental Management Program (SEMP)." It goes on to say, "Management (owner and operator) should require that the program elements are properly documented and available at field and/or office locations, as appropriate for each program element."

The standard recognizes that contractors are widely used in the offshore industry and that they should provide safe and reliable equipment and effective training to their employees. However, RP 75 does state, "This recommended practice does not require contractors to develop a SEMP. However, contractors should be familiar with the operator's SEMP and should have safety and environmental policies and practices that are consistent with the operator's SEMP." Further discussion regarding contractors is provided in RP 75's Appendix A.

The following section contains an overview of the General section APR RP 75. For detailed compliance work, refer to the actual Recommended Practice, available from the American Petroleum Institute (API).

Principles

For detailed compliance work, refer to the actual Recommended Practice document API RP 75, available from the American Petroleum Institute (API).

The General Section is divided into the following parts:

- Management Program Elements;
- Principles;
- Setting Objectives and Goals;
- Communication; and
- Scope.

A fundamental principle of RP 75 is that management is responsible for the development and implementation of the Safety Management System. Management must set goals, establish timelines, set up internal communication procedures, determine accountability and — of critical importance — provide the necessary resources. The program must be written down, and it must address all twelve technical elements of the SEMS program.

The General Section requires that human factors be considered, although it does not provide any detailed guidance as to how this is to be done.

As with any management program, individuals should be assigned to manage the elements of SEMP (discussed in Chapter 5). These individuals will develop, implement and report on progress in their particular area. Management will check on progress at regular intervals through the use of the audit process discussed in the next chapter. The assigned individuals should, of course, be properly qualified to do the work, and they should receive the necessary training.

Generally some parts of the SEMP program will require the use of outside, specialist help. Management must ensure that those specialist resources are provided and that they are integrated into the overall program.

This section not only describes the activities that are covered by RP 75 but also defines the toxic substances that are used offshore and that should be part of the SEMP. They are hydrogen sulfide (H_2S), Chlorine (Cl_2) and Ammonia (NH_3). Other chemicals and materials that have the potential to create a safety or environmental hazard should also be included in the SEMP, when they are used or when they are found in the produced oil and water. These materials can include steam, hot water, heat transfer fluids, molten sulfur and naturally occurring radioactive materials (NORM).

Appendix A—Contractor selection criteria

One of the most difficult aspects of any safety management program concerns the management of contractors. One reason for this difficulty is the contractors' range in size and capability. Some contractors carry out a few, simple tasks that have little impact on the safety of the rig or platform where they work. The contracting company that provides food to the facility is one such example. The workers for such a company do need some training in issues such as emergency response, but they are not likely to have a significant impact on safety.

Other contractors, however, have much more comprehensive responsibilities—as became very obvious in the follow-up analyses to the *Deepwater Horizon* event. Moreover, some of the contractors involved in that incident—Halliburton and

Transocean, for example—are very large companies in their own right, and are responsible for many of the activities that take place offshore.

RP 75 does not require that contractors develop their own SEMP, but it does recommend that the contractors be familiar with the operator's SEMP (this recommendation becomes a requirement under SEMS, as discussed in the next chapter. Ultimately, it is the operator of a facility who is responsible for the SEMP.

The following nonexhaustive list of contractor activities covered by SEMP is provided:

- Drilling
- Workover
- Well servicing
- Construction
- Electrical
- Mechanical
- Diving
- Boat and helicopter transportation
- Painting
- Operating
- Catering/janitorial

With respect to the selection of contractors, RP 75 suggests the use of another API standard, RP 76, *Contractor Safety Management for Oil and Gas Drilling and Production Operations*. RP 75 further suggests that contractors submit performance information in the following areas:

- A copy of the contractor's written safety and environmental policies and practices.
- Commitment by the contractor to adhere to the requirements of RP 75.
- Recordable injury and illness experience for the previous 3 years.
- Information regarding the initial employee safety orientation.
- A description of the contractor's various safety programs.
- A description of the training provided to employees.
- A description of the contractor's short-service employee training program.
- A description of the contractor's involvement in industry affairs.

RP 75 does not require contractors to develop their own SEMP—although most of the larger contract companies will probably have done so. However, contractors should be familiar with the SEMP of the owner/operator with whom they are working. If the contractor company does have its own SEMP, then a bridging document will be needed to align the operator and contractor programs. The development of such a document is discussed on page 119.

Smaller contractors may elect to implement only those elements of a SEMP that are pertinent to their own operations and activities. They will simply follow the operator's SEMP.

Appendix B—Industry codes, practices, and standards

Appendix B in RP 75 provides a list of reference documents that have achieved substantial acceptance within industry and government bodies. They are shown here in Table 3.3. Most are from the API, but other organizations, such as the ASME, are also referenced.

Three of the references, each of which belongs in the RP 14 Series, are particularly important. They are briefly discussed below.

RP 14C

The full title of this standard is *Analysis, Design, Installation and Testing of Basic Surface Safety Systems on Offshore Production Platforms*. It specifies the requirements for the analysis, design, installation, and testing of surface safety systems for offshore production platforms (not drilling or exploration facilities). Although it is a *Recommended* Practice, RP 14C had been incorporated as a legal requirement by the MMS prior to the publication of the SEMS rule (Chapter 4).

This practice is built around the idea that if hydrocarbons can remain contained in the system of pipes and vessels, then a serious process-related accident is unlikely to occur. This goal is achieved by identifying those process hazards that could lead to a release, and then installing two independent protective devices for each detectable event. The two levels of protection should be independent of, and in addition to, the control devices used in normal process operation. In general, the two levels should be provided by functionally different types of safety devices for a wider spectrum of coverage. Two identical devices would have the same characteristics and might have the same inherent weaknesses.

A common example of two separate and independent devices concerns high-pressure protection. A high-pressure switch (PSH) is the first level of response. It detects a high pressure and initiates actions such as shutting off heat sources and stopping feed streams to the affected equipment. If these actions are insufficient to control the pressure, a second device, usually a pressure safety relief valve (PSRV) opens, and quickly vents the vessel's contents to a safe location (usually a flare).

RP 14C has been criticized as being "overkill" and "ultraconservative." It has also been challenged as being outdated in a modern world of safety instrumented systems. Such criticisms will likely be addressed as updates are issued (a new edition is currently being prepared). In the meantime, the practice does seem to be effective. Failures in safety systems are rarely, if ever, cited as being a factor in major offshore events.

RP 14C is implemented using a three-step process.

(1) Create a safety analysis table (SAT) that lists undesirable events that could affect a component such as a pressure vessel. Such events include overpressure, low pressure (vacuum), a leak, liquid overflow, high temperature, and gas blowby.

Table 3.3 RP 75 Referenced Standards
▪ Bull E2 Management of Naturally Occurring Radioactive Materials (NORM) in Oil and Gas Production
▪ Publ 510 Pressure Vessel Inspection Code: Maintenance Inspection, Rating, Repair, and Alteration
▪ Publ 521 Guide for Pressure-Relieving and Depressuring Systems
▪ Publ 2004 Inspection for Fire Protection
▪ Publ 2007 Safe Maintenance Practices in Refineries
▪ Publ 2015 Cleaning Petroleum Storage Tanks
▪ Publ 2201 Procedures for Welding or Hot Tapping on Equipment Containing Flammables
▪ Publ 2207 Preparing Tank Bottoms for Hot Work
▪ Publ 2217A Guidelines for Work in Inert Confined Spaces in the Petroleum Industry
▪ Publ 2510 Design and Construction of Liquefied Petroleum Gas (LPG) Installations
▪ Publ 2510A Fire-Protection Considerations for the Design and Operation of Liquefied Petroleum Gas (LPG) Storage Facilities
▪ RP 1107 Pipeline Maintenance Welding Practices
▪ RP 2D Operation and Maintenance of Offshore Cranes
▪ RP 4G Maintenance and Use of Drilling and Well Servicing Structures
▪ RP 76 Contractor Safety Management for Oil and Gas Drilling and Production Operations
▪ RP 500 Classification of Locations for Electrical Installations at Petroleum Facilities
▪ RP 505 Recommended Practice for Classification of Locations for Electrical Installations at Petroleum Facilities Classified as Class I, Zone 0, Zone 1 and Zone 2
▪ RP T-1 Orientation Program for Personnel Going Offshore for the First Time
▪ RP T-2 Qualification Programs for Offshore Production Personnel Who Work with Anti-Pollution Safety Devices
▪ RP T-4 Training of Offshore Personnel in Non-Operating Emergencies
▪ RP T-6 Training and Qualifications of Personnel in Well Control Equipment and Techniques for Completion and Workover Operations on Offshore Locations
▪ RP T-7 Training of Personnel in Rescue of Persons in Water
▪ Spec 2C Offshore Cranes
▪ Spec 4F Drilling and Well Servicing Structures
▪ Std 1104 Welding of Pipelines and Related Facilities

(2) Create a safety analysis checklist (SAC) that lists all recommended safety devices and that shows conditions under which particular safety devices may be excluded.

(3) Create a safety analysis function evaluation (SAFE) chart. This chart shows all process components and their required safety devices.

RP 14C also provides a standard for component identification. The first letter identifies the component type. For example, C is compressor, M is pressure vessel (ambient temperature). This can be followed by a modifier. So MAV is a metering pressure vessel. Three additional digits are then be assigned to provide a unique identification for that item and its location.

RP 14G

The full title of this standard is *Fire Prevention and Control on Open-Type Offshore Production Platforms*. It provides recommendations for minimizing the likelihood of an accidental fire, and for designing, inspecting, and maintaining fire control systems. It emphasizes the need to train personnel in fire fighting, to conduct routine drills, and to establish methods and procedures for safe evacuation. The fire control systems discussed in this RP are intended to provide an early response to incipient fires to prevent their growth. They provide a base line and are not intended to preclude the application of more extensive practices to meet special situations or the substitution of other systems, which will provide an equivalent or greater level of protection.

This publication is applicable to fixed open-type offshore production platforms, which are generally installed in moderate climates and which have sufficient natural ventilation to minimize the accumulation of vapors. Enclosed areas, such as quarters buildings and equipment enclosures, normally installed on this type platform, are also addressed. Totally enclosed platforms such as those that are installed when weather conditions are very severe are outside the scope of this RP.

RP 14J

First published in 1993, API RP 14J—*Recommended Practice for Design and Hazards Analysis for Offshore Production Facilities*—is one document containing useful procedures and guidelines for planning, designing, and arranging offshore production facilities, and for performing a hazards analysis on open-type offshore production facilities (like RP 14C, the focus is on production facilities). It covers design concepts, hazard mitigation, personnel evacuation, equipment arrangements, and hazards analysis. A checklist is provided.

The standard is intended to bring together a brief description of basic hazards analysis procedures for offshore production facilities in one place. This RP discusses several procedures that could be used to perform a hazards analysis, and it presents minimum requirements for process safety information and hazards analysis, which can be used for satisfying the requirements of API RP 75.

Some of the special offshore hazards that the standard considers are listed below.

(1) Spatial limitations that may cause potential ignition sources being installed in or near production equipment.

(2) Spatial limitations that may result in quarters being installed near production equipment, pipeline/flow line risers, fuel storage tanks, or other major fuel sources.

(3) The inherent fire hazard presented by the release of flammable liquids or vapors, whether during normal operations or as a result of any unusual or abnormal condition.

(4) The severe marine environment, including corrosion, remoteness/isolation, and weather (wind, wave and current and ice).

(5) High-temperature and high-pressure fluids, hot surfaces, and rotating equipment located in or near operating areas.

(6) The handling of hydrocarbons over water.

(7) Large inventories of hydrocarbons from wells/reservoirs and pipelines connected to or crossing a producing platform.

(8) Storage and handling of hazardous chemicals.

(9) Potential H_2S releases.

Appendix C—References

This appendix provides detail for three public domain references; they are all to do with vapor cloud explosions.

Appendix D—Definitions

A list of definitions for the terms used in RP 75 is provided. The terms are:

- Contractor
- Critical equipment
- Facility
- Flag state
- Hazards analysis
- Mobile offshore drilling unit (MODU)
- Mobile offshore unit (MOU)
- Operator
- Owner
- Process
- Simultaneous operations (see page 163)
- Uncontrolled release
- Definitions
- Performance measures

An updated list of definitions is available from the IADC (International Association of Drilling Contractors) web site.

Appendix E—Performance measures

This appendix provides definitions for many of the terms used within the petroleum industry. Having consistent definitions allows for different companies to compare the results of their safety programs with industry norms. The appendix is divided into two sections: E.1—Definitions and E.2—Safety Performance Measures.

SEMP AND PSM

There are strong similarities between the elements of SEMP and those of OSHA's Process Safety Management (PSM) standard, as can be seen from Table 3.4.

Some of the topics, such as "Training" are identical, others, such as "Investigation of Incidents/Incident Investigation" are virtually the same. Probably, the two biggest differences are to do with "Employee Participation" and "Contractors."

The "Employee Participation" element is not found in SEMP, although some of the requirements under the General Provisions of SEMP could be construed as covering the topic. (Also, there are some suggestions that Employee Participation may be incorporated into SEMS II.) The success of the Employee Participation element is absolutely crucial to PSM and SEMP—process safety is not a program developed and run by a team of specialists. Instead process safety involves everyone on the facility, including full-time employees, contract workers and senior management. This element is about

Table 3.4 Elements of SEMP and PSM

SEMP	PSM
(1) General	(1) Employee participation
(2) Safety and environmental information	(2) Process safety information
(3) Hazards analysis	(3) Process hazards analysis
(4) Operating procedures	(4) Operating procedures
(5) Training	(5) Training
(6) Prestart-up review	(6) Contractors
(7) Assurance of quality and mechanical integrity of equipment	(7) Prestart-up safety review
(8) Safe work practices	(8) Mechanical integrity
(9) Management of change	(9) Hot work
(10) Investigation of incidents	(10) Management of change
(11) Emergency response and control	(11) Incident investigation
(12) Audit of SEMP elements	(12) Emergency planning and response
(13) Records and documentation	(13) Compliance audits
	(14) Trade secrets

participation, not mere communication. Indeed, it is not really a separate, stand-alone management element—it should thread through all the other elements of an SMS.

The difficult and important topic of "Contractors" is identified as a separate element in the PSM standard, but not in SEMP. This difficult and important topic is discussed elsewhere in this book, particularly on page 115.

"Trade Secrets" are included in the OSHA list. Some onshore chemical plants use or manufacture chemical compounds that are highly proprietary. This element provides those companies with a means to protect those secrets, while at the same time ensuring that the chemicals in question are properly managed. Since offshore platforms do not generally handle proprietary chemicals, this topic is not needed in the SEMP list.

The final element of SEMP is "Records and Documentation." Although this element is not found in PSM, all management systems will generate a lot of information and data—all of which needs to be recorded, indexed, and made readily accessible. Even if this item is not explicitly identified, every company and facility will need a system for managing its SMS.

REFERENCES

API (American Petroleum Institute) RP 75, 2004. Development of a Safety and Environmental Management Program for Offshore Operations and Facilities.

CHAPTER

4 Safety and environmental management systems

CHAPTER CONTENTS

INTRODUCTION

This chapter describes the Safety and Environmental Management Systems (SEMS) rule that applies to the Outer Continental Shelf (OCS) of the USA. (The manner in which regulations are created and enforced for the USA offshore oil and gas industries is described in Chapter 1.) The practical implementation of a SEMS program—including timing, cost, and schedule development—is discussed in Chapter 5.

In the years prior to the Deepwater Horizon catastrophe, the Minerals Management Service (MMS)—which is now BOEMRE—had been in the process of developing a new rule whose full title was *Oil and Gas and Sulphur Operations in the Outer Continental*

Ian Sutton: Offshore Safety Management. DOI: 10.1016/B978-1-4377-3524-6.00004-6

Copyright © 2012 Ian Sutton. Published by Elsevier Inc. All rights reserved.

Shelf—Safety and Environmental Management Systems (SEMS). It is Subpart S of 30 CFR Part 250. One of the consequences of that accident was that the agency radically increased the scope and speed of implementation of the rule.

The updated rule was published by the BOEMRE in October 15, 2010, with an effective date of November 15, 2010. Covered facilities were given one year in which to comply, i.e. their programs should be finalized by November 15, 2011. It should be noted that nothing actually happens on that date—it is not like a tax document that must be mailed in on or before the deadline. However, once November 15, 2011, has passed, owner/operators are subject to full SEMS audits and incident investigations.

The rule covers all oil and gas facilities on the OCS in US waters, and is based on the Safety and Environmental Management Program (SEMP) standard that was discussed in the Chapter 3. (The single letter change—from SEM'P' to SEM'S'—is significant.) SEMS is a *system* in which the parts link to one another.

FROM SEMP TO SEMS

With the promulgation of the SEMS rule, BOEMRE has transformed a Recommended Practice (RP 75) into a legal requirement (the word "must" occurs approximately 293 times in the SEMS rule). However, BOEMRE has gone further—it has added many more requirements. In their words,

> *In addition, BOEMRE is highlighting certain requirements from API RP 75 and further describing those requirements in the regulatory text to clarify compliance requirements.*

Use of the word "clarify" in this context is more than a little disingenuous. In fact, the agency has expanded the scope of SEMP considerably. For example, Section 5.2 of SEMP describes the requirements for operating procedures. It is quoted in Figure 4.1.

Figure 4.2 shows the additional requirements regarding operating procedures that were added by BOEMRE.

There are 295 words in Figure 4.1, and 212 in Figure 4.2. Therefore, BOEMRE's additional material has expanded the word count in the operating procedures section by more than 70%. Moreover, the additional material is more detailed. For example, specific circumstances are added to the general topic of emergency operations. The additional BOEMRE material makes the rule more prescriptive than the original SEMP standard.

BOEMRE is also carrying over some of the provisions that were included in the proposed (four-elements-only) rule. The following seven specific topics are carried over:

(1) Recordkeeping and documentation.

(2) Clarification of the differences between hazards analysis (facility level) and job safety analysis (JSA) (task level).

(3) Procedures to verify that contractors are conducting their activities in accordance with their skills and knowledge.

(4) An independent third-party of qualified designated personnel must conduct all audits.

5.2 CONTENT OF OPERATING PROCEDURES

Written procedures should include the following:

a. The job title and reporting relationship of the person or persons responsible for each of the facility's operating areas.
b. Instructions for the sound operation of each facility that are consistent with the safety and environmental information including, as appropriate: startup, normal operations, temporary operations, simultaneous operations, emergency shutdown and isolation, and normal shutdown.
    1. Refer to API 14J, *Recommended Practice for Design and Hazards Analysis for Offshore Production* Facilities (latest edition), for information on startup, normal operations, and shutdown of production facilities.
    2. Refer to MODU Operations Manual developed in conformance with flag State requirements and/or the IMO MODU code for information on routine operations and operating limits on mobile offshore drilling units.
c. The operating limits resulting from the information specific in Section 2 and, where safety and environmental considerations are present, a description of the following:
    1. The safety and environmental consequences of deviation outside the operating limit envelope.
    2. The steps required to correct or avoid a deviation from the operating limits.
d. Environmental and occupational safety and health considerations, including the following:
    1. The special precautions required to prevent environmental damage and personnel exposure, including engineering controls and personal protective equipment.
    2. The control measures to be taken if physical contact or airborne exposure occurs.
    3. Any special or unique hazards.
    4. Continuous and periodic discharge of hydrocarbon materials, contaminants, or undesired by-products into the environment is restricted by governmental limitations. These discharge limitations represent the degree of effluent reduction attainable by application of the best practicable control technology. Written guidance should be provided for facility operating personnel and contractors governing the disposal of materials within terms of the applicable permits.
    5. Any lease or concession stipulations established by the recognized governmental authority.

FIGURE 4.1

Section 5.2 of SEMP.

(5) Audit documentation must be submitted to BOEMRE.

(6) Other documentation to be made available to BOEMRE upon request.

(7) Form MMS-131 (page 231).

FIRST VERSION OF SEMS

A draft version of SEMS was introduced in 2006. In order to determine its scope the MMS analyzed their own incident data, totaling 1950 incidents, and determined that 4 of the 12 elements of SEMP contributed the most to the incidents. The elements that they selected and the data that they used are shown in Table 4.1.

The MMS held a public hearing on the proposed rule in New Orleans in September 2009. At that hearing, representatives from industry challenged the need for the new

(a) You must develop and implement written operating procedures that provide instructions for conducting safe and environmentally sound activities involved in each operation addressed in your SEMS program. These procedures must include the job title and reporting relationship of the person or persons responsible for each of the facility's operating areas and address the following:

(1) Initial startup;
(2) Normal operations;
(3) All emergency operations (including but not limited to medical evacuations, weather-related evacuations and emergency shutdown operations);
(4) Normal shutdown;
(5) Startup following a turnaround, or after an emergency shutdown;
(6) Bypassing and flagging out-of service equipment;
(7) Safety and environmental consequences of deviating from your equipment operating limits and steps required to correct or avoid this deviation;
(8) Properties of, and hazards presented by, the chemicals used in the operations;
(9) Precautions you will take to prevent the exposure of chemicals used in your operations to personnel and the environment. The precautions must include control technology, personal protective equipment, and measures to be taken if physical contact or airborne exposure occurs;
(10) Raw materials used in your operations and the quality control procedures you used in purchasing these raw materials;
(11) Control of hazardous chemical inventory; and
(12) Impacts to the human and marine environment identified through your hazards analysis.

FIGURE 4.2

BOEMRE's addition requirements regarding operating procedures.

Table 4.1 Incident Analysis: Contributing Factors.

SEMS Element	Number of Incidents	%
Mechanical integrity	726	38
Operating procedures	609	31
Hazards analysis	412	21
Management of change	203	10

rule (Parker, 2009). Their response focused on three areas: ongoing favorable safety trends, contractor management, and the distinction between Job Hazard Analyses (JHAs) and Job Safety Analyses (JSAs).

With regard to safety trends, charts such as that shown in Chapter 1 (which was developed from the MMS's own data) were used to demonstrate that progress in both safety and environmental performance were very good. It was also pointed out that the offshore oil and gas industry has a safety record that is better than that of industry overall. For these reasons it was argued that there was no justification for bringing out a new regulation.

The comments about contractors were based on the difficulties that can occur when an owner/operator tries to align its program with that of a contractor. The commenters felt that the proposed rule had not adequately addressed this problem. The MMS personnel who were present at the meeting did not respond directly to industry's comments, and made no further announcement regarding the development of SEMS in the following months.

And then came Deepwater Horizon.

The event had two immediate consequences with regard to SEMS. First, the rule was quickly finalized, and issued in October 2010 without any further hearings. Second, the MMS (by now rebadged as BOEMRE) increased the scope of SEMS by including all 12 elements of SEMP within the rule, not just the four shown in Table 4.1.

FINAL RULE

At the heart of the SEMS rule lies the following BOEMRE statement:

> *BOEMRE is incorporating by reference, and making mandatory, the American Petroleum Institute's Recommended Practice for Development of a Safety and Environmental Management Program for Offshore Operations and Facilities (API RP 75), Third Edition, May 2004, reaffirmed May 2008. This recommended practice, including its appendices, constitutes a complete Safety and Environmental Management System (SEMS) program.*

Simply put, all of RP 75 is now the law of the land. The above quotation also provides a basis for the statement that some companies have made on the lines of, "We are not concerned about SEMS because we already have an effective SEMP." Such statements are, misleading—SEMS is considerably greater in scope than SEMP.

As the final SEMS rule was being developed during mid-2010 there was considerable discussion as to whether it would include a safety case requirement. In the end it did not. It should, however, be noted that a company that has developed a modern safety management system in order to meet the requirements of SEMP and SEMS has also, in large part, prepared most of the analyses that contribute toward a safety case. Indeed, at a Chemical Safety Board public hearing in December 2010 (CSB, 2010) a comment was made that "SEMS is a safety case." Although this statement may not be literally true, there is certainly a large amount of overlap between the two systems, and they do appear to be converging, as discussed in Chapter 8.

ORGANIZATION OF THE RULE

The SEMS rule is difficult to follow. It has neither a table of contents, nor an index. The most important material—that to do with the requirements of the program elements—is located almost at the back, and is hard to find.

Table 4.2 provides a general idea of the structure of the rule. The table has three columns. The first provides the page numbers as found in the published document. The second column provides a summary as to the material that can be found on those pages, and the third column provides some discursive material.

Table 4.2 Structure of the SEMS Rule.

Pages	Material Provided	Discussion
63610	The first page provides an introduction, a discussion of BOEMRE, effective dates, and contact information. This section also states the philosophy to do with incorporating SEMP. An explanation of the original proposed four-element rule is provided. A discussion as to the role of operators and contractors is provided. This discussion is important. Further details are provided on page	The key sentence is, "It is the intent of this rule to hold the operator accountable for the overall safety of the offshore facility, including ensuring that contractors and subcontractors have safety policies and procedures in place that support the implementation of the operator's SEMS program and align with the principles of managing safety set forth in API RP 75."
63610	"Nothing in this final rule shall affect the Coast Guard's authority and jurisdiction over vessels and offshore facilities."	In practice, oversight of MODUs may be an issue.
63610-63611	The elements of SEMP are listed, along with the corresponding SEMS paragraph number.	Each numbered item has the phrase "with additional clarification" tagged on to it. The use of the word "clarification" is somewhat disingenuous; BOEMRE is actually adding a fair amount of material. Hence, SEMS > SEMP.
63611	Seven additional elements ranging from recordkeeping to use of Form MMS-131 are listed here. These are in addition to the elements of SEMP.	Once more, SEMS > SEMP.
63611	Some estimates as to the cost of a SEMS program are provided.	These estimates are likely to be much too low. Further discussion to do with the cost of SEMS is provided in Chapter 5.
63611	There is discussion as to the enforcement mechanisms that BOEMRE can use.	
63611-63613	A preamble to the rulemaking comments is provided.	The comments provide insights into how BOEMRE was thinking when it wrote the rule.
63613-63618	Responses to general comments are provided on these pages.	

Table 4.2 Structure of the SEMS Rule—*Cont'd*

Pages	Material Provided	Discussion
63618-63632	Responses to comments on the individual elements are discussed in these pages.	The comments are not discussed here. Instead they are referred to in the discussion to do with each element.
63633-63636	Instructions to do with filling out Form MMS-131 are provided here.	
63636	The specific edition of the SEMP standard (reaffirmed May 2008) to be used is identified.	
63636	The need for a JSA is identified.	
63636	A single hazards analysis for simple and multiple similar facilities is allowed.	
63637-63638	A discussion of accident panel investigations is provided.	The discussion shows how the four top elements shown in Table 4.1 were identified.
63638	The role of independent third-party auditors is described, along with the qualifications that these people will need to have.	BOEMRE state that they are more likely to participate in an audit if the auditor comes from the contractor company that developed the SEMS program.
	BOEMRE states that all costs of audits are to be paid by the operator.	
63639	A discussion on contractors is provided.	
63639-63644	Further discussion to do with the cost of SEMS is provided in these pages. Also shown are the definitions for high, moderate, and low activity operators.	It is not always clear if BOEMRE has accounted for the four separate costs associated with implementing SEMS: (1) The cost of designing and implementing the program. (2) The cost of maintaining the program. (3) The cost of audits. (4) The cost of implementing findings from activities such as hazards analyses, audits, and MOC results.

63644-63645	This section of the rule is entitled *Benefits of SEMS*. The agency's key justification for having a SEMS program is summarized in the following sentence, "… while it is difficult to quantify with any degree of certainty the human safety and environmental benefits of a comprehensive SEMS program, the financial burden estimated for developing and managing a SEMS program is minor compared to the costs associated with major accidents."	When unpacked, this sentence raises three issues—all of which are discussed elsewhere in this book. First the agency recognizes that there are no accurate metrics for measuring process safety results, and there is no way of knowing with absolute confidence as to how well industry in doing, and how much progress it is making. The second issue raised by the BOEMRE sentence is a recognition that it is difficult to estimate the cost of implementing and maintaining a SEMS program. Finally, BOEMRE, like most other agencies and companies, skates around the issue of acceptable risk. They do not provide an objective value for ALARP.
63645	References to other rules, such as those to do with small businesses, is provided on this page.	
63646-63648	Estimates as to the hours required for reporting and recordkeeping for different activity-level operators are provided on these pages.	
63648-63649	Further discussion to do with other rules and standards is provided.	
63649-83654	These pages provide the actual rule itself. It is organized into paragraphs, each of which starts with the number §250.	The details of each of these paragraphs are described and discussed at length in the remainder of this chapter.

SCOPE

The scope of SEMS is broad, as shown in the following quotation from the rule:

> *[...] your SEMS program identifies, addresses and manages safety, environmental hazards, and impacts during design, construction, start-up, operation, inspection, and maintenance of all new and existing facilities, including mobile offshore drilling units (MODU) while under BOEMRE jurisdiction and Department of Interior (DOI) regulated pipelines.*

At the 2009 public hearing, industry representatives noted that MODUs and other offshore vessels fell under the jurisdiction of the coast guard. Therefore, they claimed, the MMS, as it then was, lacked the authority to require MODU operators to implement a SEMS program.

The hazards analysis section contains additional information as to what constitutes a covered facility. It states:

> *For this subpart, facilities include all types of offshore structures permanently or temporarily attached to the seabed (i.e. mobile offshore drilling units; floating production systems; floating production, storage and offloading facilities; tension-leg platforms; and spars) used for exploration, development, production, and transportation activities for oil, gas, or sulphur from areas leased in the OCS [Outer Continental Shelf]. Facilities also include DOI regulated pipelines.*

The rule runs to 44 pages. Approximately six of those pages cover the technical requirements of SEMS. Other pages discuss administrative and legal issues; there are also about 17 pages devoted to public comments, and BOEMRE's response to those comments.

COMPLIANCE

No matter how much goodwill is expressed between regulators and the companies that they work with, at the end of the day a basically adversarial relationship exists. On a day-to-day basis the agencies and operators may work well together, but, should it be necessary, the agency can exert the full force of the law in order to make sure that its decisions are carried out. With regard to offshore operations, BOEMRE can, if violations are serious enough (in its opinion), shut-in the platform or drilling rig.

Section §250.1927 of the rule discusses potential penalties for failing to meet the SEMS requirements. It states:

> *If BOEMRE determines that your SEMS program is not in compliance with this subpart we may initiate one or more of the following enforcement actions:*
>
> *(a) Issue an Incident(s) of Noncompliance;*
>
> *(b) Assess civil penalties; or*
>
> *(c) Initiate probationary or disqualification procedures from serving as an OCS operator.*

The first action that the agency can take is to issue an incident of noncompliance—usually referred to as an INC. The following quotation is taken from the BOEMRE website:

> *Upon detecting a violation, the Bureau Inspector issues an Incident of Noncompliance (INC) to the operator and uses one of two main enforcement actions (warning or shut-in), depending on the severity of the violation. If the violation is not severe or threatening, a warning INC is issued. The warning INC must be corrected within a reasonable amount of time specified on the INC. The shut-in INC may be for a single component (a portion of the facility) or the entire facility. The violation must be corrected before the operator is allowed to continue the activity in question.*

If the operator has identified the areas of noncompliance during a routine audit and has also prepared a schedule for their correction, then there is a reduced chance that an INC will be issued.

If noncompliance could pose actual harm or a threat to the human or marine environment then the agency can issue civil penalties as well as ordering a shut-in of the facility. In extreme cases, criminal action may be brought.

CONTRACTORS

One of the biggest difficulties with the implementation of SEMS, indeed of any safety management system, is defining and managing the relationship between contractors and the operating companies that they work for. (Some companies refer to their long-term contractors as "business partners.") The rule makes it clear that, from a legal point of view, the responsibility for implementing a SEMS program lies with the operator. It is his or her responsibility to make sure that the contractors on the rig or platform are informed as to the SEMS requirements, and that the rule's requirements are properly addressed. BOEMRE cannot directly issue a noncompliance notice, levy a fine, or ban a contractor. It can do all of those things to a lessee or operator who does not ensure contractors working on its behalf are compliant with regulations.

The topic of contractor management is not called out explicitly as one of the elements in Table 4.2, although it is described in Attachment A of the SEMP standard (and is discussed in Chapter 3).

Reports from the early SEMS pilot audits suggest that the BOEMRE inspectors will be paying considerable attention to the training and knowledge of contractors (but, strictly speaking, not the competence of individuals). They are particularly interested in ensuring the quality of training in situations where there is considerable turnover. They may also require that any contractor at the facility be able to describe the SEMS program, as it affects his or her work, without referring to notes.

The OSHA regulation

In the previous chapter it was noted that one of the biggest differences between SEMP and the Occupational Safety & Health Administration (OSHA) process safety management (PSM) standard was in regards to contractors. OSHA has a section explicitly related to the topic, SEMP does not. The OSHA regulation and matching guidance is provided in Figures 4.3 and 4.4.

(1) Application. This paragraph applies to contractors performing maintenance or repair, turnaround, major renovation, or specialty work on or adjacent to a covered process. It does not apply to contractors providing incidental services which do not influence process safety, such as janitorial work, food and drink services, laundry, delivery or other supply services.

(2) Employer responsibilities.

(i) The employer, when selecting a contractor, shall obtain and evaluate information regarding the contract employer's safety performance and programs.

(ii) The employer shall inform contract employers of the known potential fire, explosion, or toxic release hazards related to the contractor's work and the process.

(iii) The employer shall explain to contract employers the applicable provisions of the emergency action plan required by paragraph (n) of this section.

(iv) The employer shall develop and implement safe work practices consistent with paragraph (f)(4) of this section, to control the entrance, presence and exit of contract employers and contract employees in covered process areas.

(v) The employer shall periodically evaluate the performance of contract employers in fulfilling their obligations as specified in paragraph (h)(3) of this section.

(vi) The employer shall maintain a contract employee injury and illness log related to the contractor's work in process areas.

(3) Contract employer responsibilities.

(i) The contract employer shall assure that each contract employee is trained in the work practices necessary to safely perform his/her job.

(ii) The contract employer shall assure that each contract employee is instructed in the known potential fire, explosion, or toxic release hazards related to his/her job and the process, and the applicable provisions of the emergency action plan.

(iii) The contract employer shall document that each contract employee has received and understood the training required by this paragraph. The contract employer shall prepare a record which contains the identity of the contract employee, the date of training, and the means used to verify that the employee understood the training.

(iv) The contract employer shall assure that each contract employee follows the safety rules of the facility including the safe work practices required by paragraph (f)(4) of this section.

(v) The contract employer shall advise the employee of any unique hazards presented by the contract employer's work, or of any hazards found by the contract employer's work.

FIGURE 4.3

OSHA regulation.

Employers who use contractors to perform work in and around processes that involve highly hazardous chemicals, will need to establish a screening process so that they hire and use contractors who accomplish the desired job tasks without compromising the safety and health of employees at a facility. For contractors, whose safety performance on the job is not known to the hiring employer, the employer will need to obtain information on injury and illness rates and experience and should obtain contractor references.

Additionally, the employer must assure that the contractor has the appropriate job skills, knowledge and certifications (such as for pressure vessel welders). Contractor work methods and experiences should be evaluated. For example, does the contractor conducting demolition work swing loads over operating processes or does the contractor avoid such hazards?

Maintaining a site injury and illness log for contractors is another method employers must use to track and maintain current knowledge of work activities involving contract employees working on or adjacent to covered processes. Injury and illness logs of both the employer's employees and contract employees allow an employer to have full knowledge of process injury and illness experience. This log will also contain information which will be of use to those auditing process safety management compliance and those involved in incident investigations.

Contract employees must perform their work safely. Considering that contractors often perform very specialized and potentially hazardous tasks such as confined space entry activities and nonroutine repair activities it is quite important that their activities be controlled while they are working on or near a covered process. A permit system or work authorization system for these activities would also be helpful to all affected employers. The use of a work authorization system keeps an employer informed of contract employee activities, and as a benefit the employer will have better coordination and more management control over the work being performed in the process area.

A well run and well maintained process where employee safety is fully recognized will benefit all of those who work in the facility whether they be contract employees or employees of the owner.

FIGURE 4.4

OSHA guidance for the regulations.

Although the materials in Figures 4.3 and 4.4 do not have any legal authority offshore, it is useful to review them to see if they provide guidance for those working under the SEMS rule.

Types of contractor

The management of contractors is important. Frequently, contract workers are involved in activities that have a high potential for injury, loss of life, environmental releases, and economic loss. In most cases, the contractor company is on site for a relatively short period of time. This means that the contractor's managers, supervisors, and workers are not likely to be completely familiar with the operating company's way of doing things.

Contract companies

Contract companies vary enormously in size. Some of them are small organizations that are on board just to carry out tasks that are limited in scope and number. However, other contractors—as was seen on Deepwater Horizon—are not only large, but also they are responsible for much of the high-risk work that is carried out on rigs and drill rigs. Clearly, there cannot be a "one-size-fits-all" contractor management program in such circumstances.

Contract workers

Some contract workers are present at a facility over a long period of time and function almost as if they are full-time workers. It is likely that these contract workers will be fully familiar with the operator's safety programs.

Other contract workers, however, are present on a facility only for a short time, but may be asked to perform high-risk activities such as vessel entry of working with high-voltage electrical equipment. In situations such as this, the operator has two tasks. First he has to establish that the contract worker has adequate general training for the type of work that he or she is performing. Second, the operator has to make sure that the temporary contract worker has received sufficient training in the operations at the particular facility, where the work is to be carried out. This validation may involve the use of bridging documents.

Design companies

The mechanical integrity part of the rule—*Assurance of Quality and Mechanical Integrity of Equipment*—can involve design companies because they are the ones who are making fundamental engineering decisions. However, BOEMRE always has to work through the lessee/operator; the agency has no legal authority to enter a contractor's workplace. They can, however, enter the offices of the lessee. If they determine that the contractor's work is not satisfactory they can issue a noncompliance notice or fine on the lessee.

Contractors are also required to follow the many prescriptive regulations that are in place. The design may also be subject to a third-party review by a Certified Verification Authority (CVA) that has been approved by the BOEMRE.

SEMS requirements

In addition to the material provided in Appendix E of SEMP, some discussion to do with the management of contractors is provided on page 63639 of the SEMS rule. It is repeated in Table 4.3. The first column shows the BOEMRE requirement, the second column provides some discussion as to what needs to be done to meet that requirement.

Bridging documents

Contractors do not work in isolation—they have organizational and physical interfaces with the operator of the offshore facility and with other contractors. Since each company will have its own Safety Management System a bridging document is needed. This document is essentially a map that links the relevant sections of the different standards. So, if the operator's plan calls for start-up procedures, then the bridging document would show where the same information is located in the contractors' documents.

Table 4.3 SEMS Contractor Requirements.

BOEMRE Requirement	Discussion
This final rule requires operators to verify that their contractors can perform their assigned duties.	This sentence places a considerable responsibility on the operating company. Even if the contracting company has been accepted for work in general, there still needs to be a validation that the workers are qualified to execute the specific tasks to which they have been assigned.
The operator is responsible for ensuring that *all* [emphasis in the original] contractors and subcontractors have safety policies and procedures in place that support the implementation of the SEMS program and align with the principles of managing safety set forth in API RP 75.	Once more, the operator's responsibility is onerous. Considerable effort will be needed to ensure that the policies and procedures for all contractors align with the SEMS program. Given that there will often be differences between the operator's and contractor's programs, it will be necessary to write bridging documents to ensure that the differences do not cause violation of the SEMS rule. The operator will have to show explicitly how his SEMS program, the bridging documents, and the contractor's safety management system address the requirements of SEMS and RP 75. This is not a trivial task.
The operator must inform contractors of any known hazards on the facility that are related to the contractor's work. This applies to contractors performing maintenance or repair, turnaround, major renovation, or specialty work on or adjacent to a covered process.	Generally, this requirement will be addressed when JSAs are being conducted and when permits to work are being written. However, the contract workers may be required to participate in other activities, such as prestart-up reviews.

The flow of information can be in either direction. For example, a large operator may hire a painting contractor. Most of the information as to how to conduct safety activities will be from the operator to the contractor, as shown in Figure 4.5.

But the contractor may have some special information to do with the paint being used, or how to control a fire should it ignite. In this case, the information would flow from the contractor to the operator.

Before work starts it will be necessary for management of both the operator and the contractor to compare their safety management plans. They will need to ensure that the contractor's program is aligned with that of the operator and that it meets the requirements of the SEMS rule.

Given that there are about 130 operators in the Gulf of Mexico alone and over 10,000 contractors (of all types), the number of potential bridging documents over a million—obviously a totally unmanageable workload. One way of getting around this is for each company to map its program to the elements of SEMS, as shown in

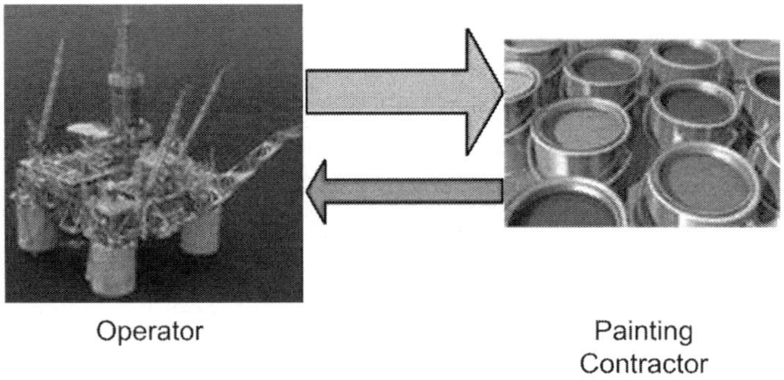

FIGURE 4.5

Most safety information will flow from the operator to the contractor.

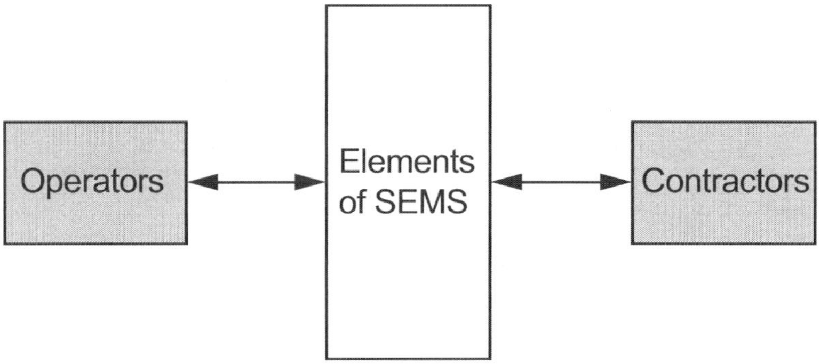

FIGURE 4.6

Mapping elements of SEMS.

Figure 4.6. The company-to-company bridging would go through this central structure. Then the number of documents goes down from 1,500,000 to about 10,000, i.e. one for each contractor and a few for the operators.

The arrows in Figure 4.6 go in both directions. Both contractors and operators receive and provide information to one another. In general, the flow of information will be from the operator to the contractor, but exceptions can occur. For example, if the contractor is a painting company, he can provide specialist information about the chemical composition of the paints and solvents that are being used.

The essential point is that all companies map their programs to the elements defined in the SEMS rule.

ELEMENTS OF SEMS

SEMS is comprised of the 12 elements of the SEMP standard described in the previous chapter. BOEMRE has incorporated these elements directly into the new rule, word for word. BOEMRE has added additional requirements for each of the elements. These requirements are available on pages 63650–63654 of the rule.

This section lists the requirements for each of these elements from both SEMP and the later BOEMRE requirements. For each element the following information is provided:

- The SEMP/SEMS rule
- Additional BOEMRE requirements
- Discussion

The actual text of API RP 75 is not quoted directly for information rights reasons. A condensed and edited version of the SEMP requirements are provided here. For detailed compliance work refer to the actual Recommended Practice.

For the first two of the above items a tabular approach is used. In the left-hand column are the words of the standard or rule *verbatim*. For editing purposes some minor changes may have been made to the original text and some of the internal cross-references have been removed. Therefore, what is provided here should not be used for formal compliance work. The second column provides context, discussion, and interpretation. The discussion section provides general information to do with that element.

This book is not the place to provide detailed guidance regarding the implementation of topics such as operating procedures. Such guidance is available in a sister book in this series, *Process Risk and Reliability Management*, details of which are provided at www.stb07.com/publications/book-process-risk-reliability-management.html.

GENERAL

The first part of the SEMP standard covers general provisions and management responsibilities. BOEMRE states that the general section of RP 75 (Chapter 3) is incorporated into SEMS. The following quotation is from the rule:

> *This General element is critical to the successful implementation of the SEMS program in API RP 75, and BOEMRE is including it by incorporating by reference the entirety of API RP 75.*

The SEMS rule

The SEMP/SEMS requirements concerning the general and management issues for owners and operators are described and discussed below. Only the first paragraph—1.1 Purpose and Objective—is discussed here. Paragraphs 1.2 through 1.4 are discussed elsewhere in this book, particularly in Chapter 3.

The General part of SEMP is divided into the following sections:

1.1. Purpose and Objective

    1.1.1. Owners and Operators

    1.1.2. Contractors

1.2. Management Program Elements and Principles

    1.2.1. Management Program Elements

    1.2.2. Principles

1.2.3. Setting Objectives and Goals

1.2.4. Communication

1.3. Scope

1.3.1. Applications

1.4. Definitions

1.5. Standards, Regulations, and References

1.5.1. Industry Codes, Practices and Standards

1.5.2. Government Codes, rules, Conventions and Regulations

1.5.3. References

Basically, the General section shows that the purpose of RP 75 is that companies should develop a SEMP that addresses both safety and environmental issues. It covers all aspects and phases of an offshore operation, including the construction, start-up, operation, inspection, and maintenance, of new, existing, or modified drilling and production facilities. General management principles to do with objectives, goals and communication are described.

This section also makes it clear that the primary responsibility for the development and application of a SEMP lies with the owner/operator. (The phrase owner/operator is often used in RP 75, suggesting that the two entities are the same. However, this is not always the case. Other terms that are used are lessee and duty holder.) This section also discusses the role of large contractors, their need for a safety program and the value of them developing a SEMP. The standard states, 'This recommended practice does not require contractors to develop a SEMP. However, contractors should be familiar with the operator's SEMP... '. Reference is made to the need for bridging documents.

The General section identifies the twelve technical elements of a SEMP, and makes it clear that all information to do with SEMP should be properly recorded. (One of the difficulties that investigators ran into regarding the Piper Alpha disaster was that key work orders were either destroyed by the fire or sank with the rig. Therefore onshore backups should be made of critical documentation.)

BOEMRE requirements

§250.1909 What are management's general responsibilities for the SEMS program?

BOEMRE Requirements	Discussion
You, through your management, must require that the program elements discussed in API RP 75 (incorporated by reference as specified in §250.198) and in this subpart are properly documented and are available at field and office locations, as appropriate for each program element. You, through your management, are	BOEMRE places the responsibility for meeting the rule with the facility management (this is similar to the safety case philosophy whereby those who create the risk should also manage that risk). The location of SEMS-related information needs some thought. Critical and time-sensitive information such as emergency response

BOEMRE Requirements	Discussion
responsible for the development, support, continued improvement, and overall success of your SEMS program. Specifically you, through your management, must:	procedures must be available on the platform or rig. Other information, such as training records, can be stored onshore (although, even then, it is possible that a SEMS auditor may require that someone working offshore be able to produce his or her training records on the spot).
(a) Establish goals and performance measures, demand accountability for implementation, and provide necessary resources for carrying out an effective SEMS program.	No specific guidance is given regarding the setting of goals, or how progress is to be measured. It is fundamentally difficult to set performance and accountability measured for a performance-based system, in which success is measured not by compliance with specific rules but by not having accidents. The provision of resources for SEMS is discussed in the next chapter.
(b) Appoint management representatives who are responsible for establishing, implementing, and maintaining an effective SEMS program.	Most companies will appoint a manager who has overall responsibility for the SEMS program. This will often be the offshore installation manager (OIM). Reporting to him or her will be a SEMS coordinator who is responsible for the day-to-day running of the program. It is likely that most of the senior managers involved in the SEMS program will not be present at the facility. They could be on another platform or they may work onshore—maybe in another part of the country. This lack of presence can cause problems when it is necessary to make "big" decisions quickly. For example, the Piper Alpha catastrophe deteriorated so badly because an OIM on another platform continued to pump oil into the conflagration. It would also appear that lack of communication between those working on the Deepwater Horizon rig and the managers onshore was a factor in that disaster. Further discussion of the organization of a SEMS program is provided in the next chapter.
(c) Designate specific management representatives who are responsible for reporting to management on the performance of the SEMS program.	Responsibilities for the specific elements will usually be delegated to the appropriate departmental manager. For example, the maintenance manager will be responsible for mechanical integrity. Other functions, such as incident investigation, may be considered to be a corporate function.

BOEMRE Requirements	Discussion
(d) At intervals specified in the SEMS program and at least annually, review the SEMS program to determine if it continues to be suitable, adequate, and effective (by addressing the possible need for changes to policy, objectives, and other elements of the program in light of program audit results, changing circumstances, and the commitment to continual improvement), and document the observations, conclusions, and recommendations of that review.	This paragraph is praiseworthy in concept, but difficult to nail down. The word "effective" is rather vague. The only way that a performance-based system can be considered to be effective is not to have accidents—yet no company, regardless of the quality of its safety management systems can make that claim. One way of meeting this requirement is for management to bring in someone from another company or division to evaluate progress with the SEMS program, and to recommend changes, if needed.
(e) Develop and endorse a written description of your safety and environmental policies and organizational structure that define responsibilities, authorities, and lines of communication required to implement the SEMS program.	In Chapter 5, it is suggested that companies and facilities develop a SEMS manual. The organization of the program can be included in that manual. It is likely that the most serious organizational difficulties will occur at the interfaces with other companies, particularly contractors and vendors. For larger companies, it is also important to work out the organizational links between head office and the field operations.
(f) Utilize personnel with expertise in identifying safety hazards, environmental impacts, optimizing operations, developing safe work practices, developing training programs, and investigating incidents.	Of all the requirements that a SEMS program places on companies, this could be the toughest. The employees who possess the necessary expertise are the ones who are needed to design and operate the platforms and rigs. Finding employees or outside experts with the knowledge and experience to address the items listed here is difficult, and the competition for such personnel is likely to increase as more and more companies implement SEMS. In many parts of the rule, BOEMRE talks about the need to use independent, third-party auditors. These people will also need to be experienced in offshore operations and will draw from the same pool of skilled personnel.
(g) Ensure that facilities are designed, constructed, maintained, monitored, and operated in a manner compatible with applicable industry codes, consensus standards, and generally accepted practice as well as in compliance with all applicable governmental regulations.	The SEMP standard lists most of the major standards that are used offshore. Most of them are issued by the American Petroleum Institute (API). Given the sheer number of rules and standards, it will be necessary to create a checklist to make sure that none of them are overlooked.

BOEMRE Requirements	Discussion
(h) Ensure that management of safety hazards and environmental impacts is an integral part of the design, construction, maintenance, operation, and monitoring of each facility.	In practice, this requirement will be mostly addressed through the use of hazards analyses.
(i) Ensure that suitably trained and qualified personnel are employed to carry out all aspects of the SEMS program.	This requirement follows on from those discussed above regarding having a sufficient number of qualified people to develop, run, and maintain the SEMS program.
(j) Ensure that the SEMS program is maintained and kept up-to-date by means of periodic audits.	Audits should be conducted by personnel who are not regular workers at the facility and who are not in line management. This gives them the freedom to state what they find without fear of being overridden. The topic of auditing is discussed on page 173.

SAFETY AND ENVIRONMENTAL INFORMATION

The first of the program elements of SEMS is called "Safety and Environmental Information." The scope of this element is actually quite a bit broader than the title indicates. Virtually all technical information, including Piping & Instrument Diagrams (P & IDs) and equipment data sheets, are incorporated into this element because these documents provide the basis for safety and environmental information.

The SEMS rule

The SEMP/SEMS requirements to do with safety and environmental information are summarized and discussed separately.

SEMS Standard	Discussion
1.1 General	
Safety and environmental information provides the technical basis for any SEMS program (Records and Documentation are discussed below). It is particularly important to have accurate and complete information for hazards analyses.	For an organization that does not have a SEMP or SEMS program at all, the first actions should be to build up the information base, particularly the P&IDs, because they are so foundational to all the other activities, such as hazards analysis, operating procedures and management of change. The level of effort needed to meet the requirements of this paragraph is substantial. As noted above, it goes beyond just 'safety information' — all the technical information that is needed to run a safe facility must

SEMS Standard	Discussion
	be collected, organized and kept up to date. On older facilities it is frequently found that the documented information does not match what is in the field. Over the years it is likely that many changes have been made and not properly recorded. Therefore a considerable amount of time and money has to be allocated toward simply matching with what is in the field against the drawings, equipment lists, instrument diagrams and data sheets. This can be an onerous activity. A problem that sometimes arises is that much of the information is stored in offices onshore, and therefore may not be easily accessible for those working offshore, particularly at night and during off-hours. Therefore it is particularly important to identify the information that could be needed quickly by the operations personnel during the off-hours, particularly if that information is crucial in responding to emergencies or severe process upsets. The same principle can be applied to other areas of the SEMS program, particularly hazards analysis. Once one unit has been analyzed, the others can be reviewed 'by difference'. It is important, however, to make sure that there really are not differences, particularly on older facilities where they may have been a good deal of 'midnight engineering'.
1.2 Process Design Information	
Process design information is information that is to be found on P&IDs (Piping and Instrument Diagrams), PFDs (Process Flow Diagrams), safe upper and lower limits, layouts, safety systems (instrument and mechanical), and Electrical One-Line Drawings. RP 75 refers to RP 14 J as a source of guidance.	The process design information provides the foundation for the hazards analyses (which can themselves provide technical information). The development of 'acceptable upper and lower limits' is both difficult and important. It is difficult because neither the designers nor the operations personnel may actually know what these values are. It is important because, if process conditions go outside the 'acceptable limits' then the operation is, by definition, unsafe. It also means that, should management choose to operate outside the safe limits envelope, then they must first initiate the Management of Change process.
1.3 Mechanical and Facilities Design Information	
	The comments made above with respect to process information apply equally to the mechanical and general facilities information, i.e. the information is hard to find and time-consuming to collect and organize. But the effort must be made – accurate and

SEMS Standard	Discussion
	complete technical information provides the foundation of all the other elements of SEMS.
The mechanical and facility design for mobile offshore units (MOUs) should conform to the applicable requirements of the flag State and classification society.	In practice, navigating through the many rules and regulations for MOUs is time-consuming and requires considerable knowledge and skill. During a design project, plenty of time should be allowed for this work.
The design should be based on the standards that were in place at the time it was designed.	The standard allows for 'grandfathering' of old equipment that was designed to a previous version of a standard or code. However, the implementation of a SEMS program does provide management of older facilities to decide on whether to upgrade old equipment to meet the latest standards, even if they are not actually required to do so. This concept applies particularly to the final paragraph – if a hazards analysis finds that some equipment is not safe, then it should be made safe, regardless of the specifics of the pertinent codes and standards (see the next section).
If the original design information is not available, then the suitability of equipment for continued service can be determined through engineering analysis.	Care has to be taken with regard to 'successful prior operating experience'. The fact that a piece of equipment has performed well in the past is no proof that it can handle process conditions outside the safe limits, unless there is evidence of such operation.
The standard states that human factors should be considered. A reference to ASTM F1166-95, *Standard Practice for Human Engineering Design for Marine Systems, Equipment, and Facilities* is provided.	Apart from the reference provided no guidance is given in either SEMS or SEMP (or in the OSHA Process Safety Management standard) as to how human factors analysis and management is to be implemented.

BOEMRE requirements

§250.1910 What safety and environmental information is required?

BOEMRE Requirements	Discussion
(a) You must require that SEMS program safety and environmental information be developed and maintained for any facility that is subject to the SEMS program.	All safety management systems are rooted in accurate technical information that is easy to access, and that it also easy to modify (under the MOC program). The development and maintenance of this information is time-consuming and expensive, but it has to be done.
(b) SEMS program safety and environmental information must include: (1) Information that provides the basis for implementing all SEMS program	Paragraph (b)(1) provides a basis for deciding which activities to carry out first. Given that many companies will not be able to meet all of the SEMS requirements within the time that

BOEMRE Requirements	Discussion
elements, including the requirements of hazard analysis (§250.1911); (2) process design information including, as appropriate, a simplified process flow diagram and acceptable upper and lower limits, where applicable, for items such as temperature, pressure, flow, and composition; and (3) mechanical design information including, as appropriate, piping and instrument diagrams; electrical area classifications; equipment arrangement drawings; design basis of the relief system; description of alarm, shutdown, and interlock systems; description of well control systems; and design basis for passive and active fire protection features and systems and emergency evacuation procedures.	they are allowed, they should develop a prioritization system through a high level hazards analysis. The process design information resides primarily in the Process Flow Diagrams (PFDs). In many cases the PFD serves double duty as the simplified P & ID. P & IDs are foundational documents; they must be complete and accurate before full process hazards analyses can be carried out. Offshore, equipment arrangement drawings—generally referred to as the General Arrangement (GA)—are critical documents because so many safety problems concern the limited space on board a typical platform or rig.

HAZARDS ANALYSIS

SEMS requires that a hazards analysis be carried out. The purpose of the analysis is to identify and evaluate the risk associated with or environmental incidents. If the risk level is judged to be too high, then either the likelihood or the consequences of the hazard should be reduced to a point where the overall risk is acceptable.

The SEMS rule

The SEMS requirements to do with hazards analysis are summarized and discussed below.

SEMS Standard	Discussion
2.1 Application	
A hazards analysis should be conducted on all covered facilities.	This paragraph raises many large and profound issues, some of which are outside the scope of this book. The first question is whether the hazards to be identified are to do with technical, process or occupational safety. In this regard, the SEMS standard is somewhat confusing. The topic of Job Safety Analysis (JSA), which is mostly to do with occupational safety, is an integral part of the standard. Yet, RP 75 is fundamentally a *management* standard, and so the focus of

SEMS Standard	Discussion
	a hazards analysis should be on process safety issues, and the avoidance of catastrophic events such as Deepwater Horizon. Technical safety issues, such as explosion modeling or noise analysis, can be considered as falling into the domain of the first SEMP element: Safety and Environmental Information. They are discussed in Chapter 7.
	The second major issue raised in this paragraph is to do with the word 'unacceptable'. If risks are to be categorized for acceptability then some objective criterion is needed. Most companies use a simple risk matrix, such as that shown in Chapter 1 to determine which hazards are acceptable, and which are not. A more sophisticated approach is to use the ALARP concept. The single sentence to do with Human Factors provides no information as to the scope of the topic or how it is to be integrated into a hazards analysis. (OSHA is equally cryptic in its PSM standard.)
2.2 Methodology	
2.2.1 Hazards Analysis Approach	
RP 75 suggests that RP 14J be considered when deciding on the methods to use for hazards analyses.	There are many well-established and proven ways of conducting a hazards analysis. At one extreme as simple and relatively unstructured methods such as a Major Hazards Analysis. At the other end of the spectrum lie highly organized methods such as Hazard and Operability Analysis (HAZOP) and Layers of Protection Analysis (LOPA). An extended discussion of these techniques is provided Chapter 3 of *Process Risk and Reliability Management.*
Guidance to do with locations that are connected to one another is provided, along with a discussion as to how to handle a facility that has many similar operations in one location.	The philosophy behind this requirement was discussed in Chapter 1, when it was noted that the Gulf of Mexico has hundreds of small, shallow water platforms that can be analyzed generically (in contrast to the need for more individual risk analyses in the North Sea). This approach allows for conducting hazards analysis 'by difference', i.e. one analysis is used as the basis for the analyses of similar facilities (Levine, 2011). It is critical that the analysis team leader carefully check that two platforms that are supposedly identical really are identical. On older facilities in

SEMS Standard	Discussion
	particular, the assumption that two units are identical is often wrong. Over the years many changes will have been made, and it is unlikely that each unit was changed in just the same manner. Topics to consider include the age of the platform, depth of water, degree of corrosion and well pressures.
2.2.2 Hazards Analysis for Mobile Offshore Units	
a. Mobile offshore units (MOUs) require flag and classification certification. This certification will generally require that a hazards analysis be carried out.	Wells are usually drilled from floating units such as drill ships, jackups, semisubmersibles and submersibles. These Mobile Offshore Drilling Units (MODUs) fall under the authority of the Coast Guard, and their safety with respect to code is reviewed by one of the class societies. The extent to which BOEMRE has jurisdiction over MODUs is a legal question. However, if those involved with the design or operation of a MODU believe that there are hazards that have not been properly identified, then a supplemental hazards analysis should be performed.
b. Site-specific hazards analyses for MOUs may be required depending on the operating conditions. The operator will supply information to do with conditions such as the condition of the sea bed and the pressure and temperature of the reservoir.	This paragraph provides examples of circumstances where additional hazards analyses should be carried out. It does not specify a methodology, but a Hazard Identification Study (HAZID) is likely to be effective.
2.3 Initial Analysis	
2.3.1 Existing Facilities	
An order of priority for conducing hazards analyses on existing facilities is required. Factors to consider include the location of living quarters, inventories of hazardous materials, simultaneous operations, the presence of toxic gases, the severity of operating conditions and any special environmental issues.	The standard requires that a defensible priority be given to the hazards analyses. In addition to the criteria shown to the left, a Major Hazards Screening organized around a risk matrix can be used to come up with a preliminary evaluation of each facility or process within the facilities. In addition to using a risk ranking, the order of analyses will be influenced by factors such as the availability of up to date drawings, the facility's incident history and the availability of qualified personnel who know the areas being evaluated.
2.3.2 New and Modified Facilities	
This section outlines issues to consider when conducting a hazards analysis.	In order to ensure formal compliance with the SEMS rule, the team leader or scribe should explicitly call out each one of these factors during the team

SEMS Standard	Discussion
	discussions, and when the report is being written. He or she should show that the requirements of each sub-paragraph have been addressed.
2.4 Periodic Analyses	
Analyses should be updated on a regular basis – usually in the 5 to 10 year range, depending on the risk associated with the facilities under review.	If the facility's Management of Change program is operating effectively, there should not be any need to update hazards analyses. The reality, however, is that a review program is needed in order to catch any changes or modifications that may have slipped through without being properly evaluated. The five year period is in line with OSHA's PSM standard.
2.5 Analysis Personnel	
RP 75 requires that hazards analyses be performed by competent personnel, with the appropriate technical skills, and that the meetings should be led by someone experience in hazards analysis.	This paragraph does not require that hazards analyses be conducted by teams. However, except for certain specialized applications such as Fault Tree Analysis, a team approach is almost always used. There are two important reasons for this. First, each person on the team is likely to have some knowledge or expertise that he or she can share with others. Second, a team can brainstorm and generate new insights and ideas. Those who are involved in the design (or operation) of a facility should not lead the analysis. The leader should be independent of the organization responsible for the design or operation of the facility. A major practical problem is that there are not enough qualified people available to industry and the regulators if a large amount of hazards analysis work is to be carried out in the short time available.
2.6 Analysis Report	
All hazards analyses should result in a report that describes what was done, and the recommendations generated.	The report phase of a hazards analysis is often poorly done. The people involved in the team meetings move on to new projects, and tend to put off the writing and the reviewing of reports on studies that are already complete. If the leader of the analysis is an outsider it is essential that he or she is provided with sufficient funds, time and access to local technical experts to make sure that a thorough report is written. The reference to qualitative assessments is important. RP 75 is not requiring that a Probabilistic Risk Analysis be carried out, which implies that neither need a Safety Case be prepared.

BOEMRE requirements

§250.1911 What criteria for hazards analyses must my SEMS program meet?

BOEMRE Requirements	Discussion
You must ensure the development and implementation of a hazards analysis (facility level) and a JSA (operations/task level) for all of your facilities.	Although the rule puts process hazards analysis and JSA in the same category, they are, in fact, quite different from one another. The first addresses process hazards, the second occupational hazards. Each approach uses different techniques to identify hazards and tends to generate different types of action item.
For this subpart, facilities include all types of offshore structures permanently or temporarily attached to the seabed (i.e., MODUs; floating production systems; floating production, storage, and offloading facilities; tension-leg platforms; and spars) used for exploration, development, production, and transportation activities for oil, gas, or sulfur from areas leased in the OCS.	The SEMS rule is comprehensive. All types of offshore operation are covered (although, as discussed above, the jurisdiction to do with MODUs may lie more with the coast guard and the class societies than with BOEMRE).
Facilities also include DOI regulated pipelines.	Pipelines that are regulated by the Department of Transportation (DOT) are excluded. Typically these pipelines are not attached to a specific platform but are general purpose transmission lines.
You must document and maintain current analyses for each operation covered by this section for the life of the operation at the facility. The analyses must be updated when an internal audit is conducted to ensure that it is consistent with the current operations on your facility. Hazards analysis requirements for simple and nearly identical facilities, such as well jackets and single-well caissons, may be fulfilled by performing a single hazards analysis which you can apply to all such facilities after you verify that any site-specific deviations are addressed in each of the elements of your SEMS program.	The principle of "hazards analysis by exception" is discussed above in the SEMP paragraph 3.2.1, as is the importance of ensuring that all the differences between specific facilities and the generic base are properly identified.
(a) Hazards Analysis (facility level) For a hazards analysis (facility level), you must perform an initial hazards analysis on each facility on or before November 15, 2011.	In addition to the general requirement for completing a SEMS program, the rule explicitly states that hazards analyses on each facility must be done before the November 2011 deadline. Given that it takes time—typically at least a month—to organize such a study, the effective deadline is a good deal earlier. Moreover, as discussed in the

BOEMRE Requirements	Discussion
	next chapter, critical technical documentation such as P & IDs and layout drawings must be ready before the hazards analysis can commence.
The hazards analysis must be appropriate to the complexity of the operation and must identify, evaluate, and manage the hazards involved in the operation.	In order to be sure that the analysis is "appropriate to the complexity of the operation" a preliminary review of the system should be carried out. The review will consider factors such as the severity of the consequences of an incident, whether previous hazards studies have been performed, and the complexity of the instrumentation and controls. Depending on the results of the review, management and the facility's process safety professionals will choose the technique to be used. Some of the more commonly used methods are described on page 149.
(1) The hazards analysis must address the following: (i) Hazards of the operation; (ii) Previous incidents related to the operation you are evaluating, including any incident in which you were issued an incident of noncompliance or a civil or criminal penalty; (iii) Control technology applicable to the operation your hazards analysis is evaluating; and (iv) A qualitative evaluation of the possible safety and health effects on employees, and potential impacts to the human and marine environments, which may result if the control technology fails. (2) The hazards analysis must be performed by a person(s) with experience in the operations being evaluated. These individuals also need to be experienced in the hazards analysis methodologies being employed. (3) You should assure that the recommendations in the hazards analysis are resolved and that the resolution is documented.	The requirement to incorporate the findings from previous incidents and Incidents of Noncompliance (INCs) could be time-consuming, both in terms of background research and the commitment of members of the SEMS team. It is assumed that paragraphs (iii) and (iv) refer to instrumentation and control, covering normal control systems, and dedicated Safety Instrumented Systems (SIS). In practical terms, both the industry and the regulators are probably going to have trouble finding sufficiently qualified people. The management of findings and recommendations is time-consuming—particularly the follow-up phase. It is important to provide sufficient resources for this activity.
(b) Job Safety Analysis	This section of the SEMS rule is not derived from SEMP, nor would it normally be considered

BOEMRE Requirements	Discussion
	part of a process safety program. JSAs typically focus on occupational safety issues. Therefore, its inclusion in SEMS is somewhat unexpected. The justification for this decision is that: *The hazards analysis section in API RP 75 addresses the job task at the facility level. Therefore, BOEMRE is requiring JSAs as part of the SEMS program.* In fact, the hazards analysis section of RP 75 does not address job tasks; as its reference to RP 14C and RP 14 J indicates. RP 75 is concerned with process and technical safety.
You must develop and implement a JSA for OCS activities identified or discussed in your SEMS program. (1) You must keep a copy of the most recent JSA (operations/task level) at the job site and it must be readily accessible to employees. (2) Your JSA must identify, analyze, and record: (i) The steps involved in performing a specific job; (ii) the existing or potential safety and health hazards associated with each step; and (iii) the recommended action(s)/procedure(s) that will eliminate or reduce these hazards and the risk of a workplace injury or illness. (3) The supervisor of the person in charge of the task must approve the JSA prior to the commencement of the work.	Given that many tasks are quite similar to one another, old JSAs will be used as go-bys for new work. Depending on the scope of the work, additional signatures, such as from the Safety Department, may be required.

Hazards analysis offshore

Process hazards analysis (PHA) techniques were developed for the onshore chemical and refining industries. These industries typically work with a much wider range of chemicals than are used offshore, and many of those chemicals are highly toxic and/or corrosive—something that is rarely a major consideration offshore. Therefore, the use of onshore-style PHA methods when analyzing the risks associated with offshore facilities can feel like something of a misfit. Where possible these techniques should be modified so that problems to do with hazardous chemicals receive less prominence, but other issues—such as dropped objects and escape and evacuation routes—are analyzed more thoroughly.

The hazards analysis should also give consideration to reliability and troubleshooting issues. The processing steps on a typical platform are usually quite simple; therefore, process optimization may not be as important to the overall economics as it would be for an onshore chemical plant or refinery. What is important, however, is operations should not be interrupted. Therefore, it is useful if the hazards analysis can be oriented to identify potential production stoppages and any appropriate follow-up troubleshooting techniques.

Hazards analysis techniques

Many books and papers that explain the various types of hazards analysis techniques are available. A very brief overview of some of the techniques that are used offshore is provided in this section.

Major hazards analysis

If a facility does not have a SEMP or SEMS program, one of the first steps in the development of such a program is to conduct a Major Hazards Analysis (MHA). The purpose of this analysis is to identify those areas of the facility that pose the greatest risk. This type of analysis, which will be qualitative in nature, will show, for example, that the risers present a greater risk than the platform's gas compressors. Therefore, the initial focus of the risk management program should be on the risers.

It will likely be conducted in the same manner as a HAZID. However, the discussion will focus on the consequences of credible hazards, particularly major safety or environmental consequences.

Hazard identification (HAZID)

A HAZID exercise is a qualitative analysis carried out by a team. Its purpose is to identify specific hazards, and then to rank them by risk. In order to carry out a HAZID, the team will have to have process and layout drawings, along with information on major equipment items and control systems.

There is no single standard concerning how a HAZID should be conducted; many facilities choose to use a "slimmed down" HAZOP approach, i.e. the analysis is conducted by a multidisciplinary team that uses the deviation guideword approach that was described above. The team focuses on identifying major hazards that could materially affect cost or schedule; the nodes are generally much bigger than in a HAZOP (typically each P & ID represents a node), and only those guidewords that are likely to be relevant are used. For example, the process analyzed during one HAZID did not have any heaters, coolers, or chemical reactions. Therefore, the guidewords "High Temperature" and "Low Temperature" were excluded from the discussion (although any team member was free to raise these guidewords if he or she saw fit). As a rule of thumb, a HAZID will take about 20% of the time for a full HAZOP of the same facility.

The hazard and operability method (HAZOP)

The HAZOP method is probably the most widely-used hazards analysis method. Even those who are not familiar with the hazards analysis process will often have heard of the term HAZOP, even if they are not really sure what it means. For example, when the PSM regulations in the USA were being promulgated in the early 1990s it was not unknown for a plant manager to say, "I know what PSM is, it's HAZOPs!" In fact, the HAZOP method is just one-seventh of one of the 14 elements of PSM listed in the OSHA standard. These managers were, however, somewhat justified in what they said because they knew that, unless they could identify hazards, they could not reduce risk, and they knew that the HAZOP technique was widely accepted. Furthermore, both regulators and legal advisors generally support use of the HAZOP technique because of its reputation and because it is so thorough. Selection of the HAZOP technique is very defensible if a company is challenged regarding its safety performance, particularly in a legal dispute.

As a result of its widespread use and acceptance, large numbers of people are now trained in the use of the HAZOP method, and many of those are also trained as leaders/facilitators. Furthermore, a HAZOP infrastructure has developed. Many consulting companies offer HAZOP facilitation services; other companies provide special-purpose software.

The basic structure of a HAZOP is to divide the unit to be analyzed into nodes. A node represents a section of the process where a significant process change takes place. For example, a node might cover the transfer of material from one vessel to another through a pump. In this case, the process change is the increase in pressure and flow that occurs across the node. Another node might include an overhead air-cooler on a distillation column. Here, the temperature and phase are the process variables that change.

Although the strength of the HAZOP method lies in its clear organization, it is important not to allow the analysis to become too rigid. If the team finds that it is talking about "Reverse Flow" even though the current guideword is "High Flow," the leader should probably let the discussion continue. If he or she were to postpone the discussion until the "right" guideword, the current thinking and creativity may be lost. On the other hand, the leader must also keep the discussion focused on the issue at hand, and should prevent too many digressions.

The steps that the HAZOP team works through are listed in Figure 4.7.

Bow-Tie analysis

A hazards analysis technique that is widely used on offshore facilities is known as bow-tie analysis (the name is based on the shape of the final analysis product, as shown in

1. Select a node, define its purpose and determine the process safe limits.
2. Select a process guideword.
3. Identify the hazards and their causes using the deviation guidewords.
4. Determine how the hazard is 'announced', i.e. how the operator knows a safe limit has been exceeded.
5. Estimate the consequences of each hazard.
6. Identify the safeguards.
7. Estimate the frequency of occurrence of the hazard.
8. Risk rank the hazard, with and without safeguards.
9. Develop findings and potential recommendations.
10. Move on to the next process guideword, or to the next node if the guideword discussion is complete.

FIGURE 4.7

Steps for the HAZOP team.

Figure 4.8). The method is used not only to identify risks, but also to assess the effectiveness of the system's safeguards.

Figure 4.8 shows the structure of a bow-tie diagram, which consists of an event tree that creates a top event, the safeguards and controls for which are analyzed with a fault tree. At the left of the diagram is the hazard, say a vessel containing hydrocarbons stored under pressure. Threats are events such as corrosion, external impact, and operating error, which could create an undesired event, in this case a release of hydrocarbons from the vessel. Between the threats and the undesired outcome are barriers such as operator training, relief valves and instrumentation.

Moving to the right side of the diagram, if the top event does occur, various recovery and preparedness measures limit its impact. In the case of the over-pressured vessel, these measures could include emergency blowdown or the triggering of a deluge system. If these protective measures do not work, a range of consequences such as fire or the release of toxic materials to the atmosphere will take place.

Examples of barriers on the initiating event side include the following:

- Trained personnel
- Equipment design
- Layout

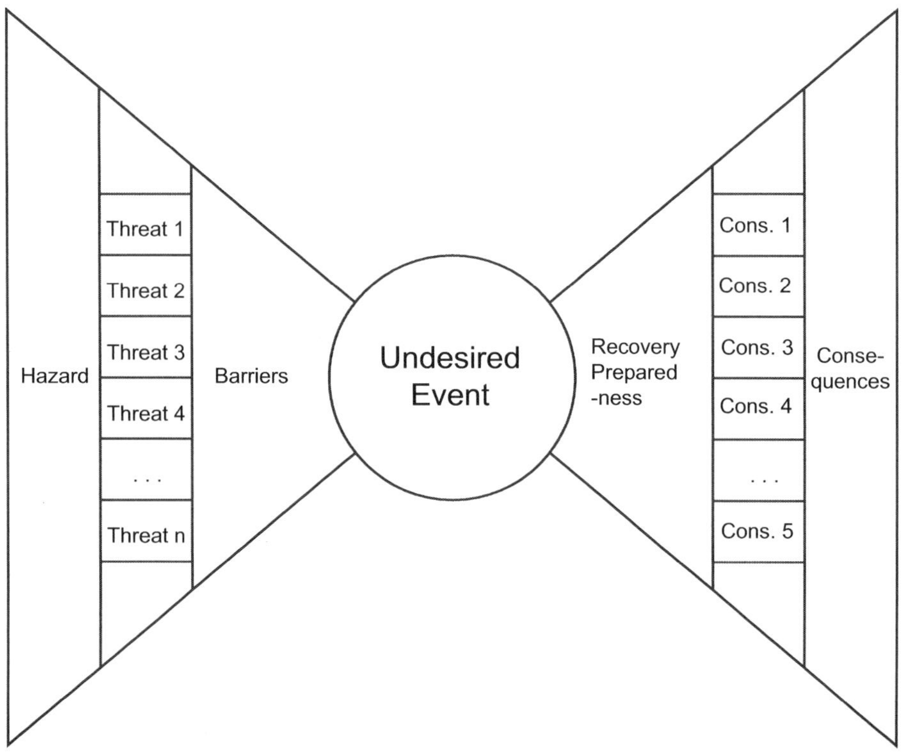

FIGURE 4.8

Bow-tie analysis.

- Inspection
- Maintenance
- Instrumentation

Examples of barriers on the mitigation side include the following:

- Well control systems
- Shutdown systems
- Detection systems
- EER

Failure modes and effects analysis

One hazards analysis technique used to analyze equipment items is Failure Modes & Effects Analysis (FMEA). The method examines the ways in which an equipment item can fail (its *failure modes*), and examines the *effects* or consequences of such failures. If the criticality of each failure is to be considered, then the method becomes a Failure Modes, Effects and Criticality (FMECA) Analysis. The consequences can be linked to safety, reliability, or environmental performance.

An FMEA is a bottom-up approach to hazards analysis. When linked with a top-down method (such as Fault Tree Analysis) a powerful synergy can ensue. The top-down method will highlight those areas which pose the greatest risk; the FMEA can then be used to investigate those areas in greater detail. Like other types of hazards analysis, an FMEA should be carried out by a team. In most cases, however, only two or three team members—who are specialists in the required fields—are involved.

Historically, the FMEA technique has been extensively used in the aerospace, automotive, electronics, and defense industries because they all require analysis of complex mechanical systems and because the failure of an equipment item can have such catastrophic consequences. The FMEA method has not been used a great deal in the process industries, partly because of a perception that its use is very time-consuming. (The same criticism is sometimes made of Fault Tree Analysis). In reality, neither the FMEA nor FTA methods need take a lot of time; it is just that the level of detail that is necessary for the analysis of say a nuclear reactor or airplane wing is much greater than that needed for a pump in a refinery because the immediate consequences of a failure impact are likely to be so much greater.

The following are components of a typical FMEA:

- Determine the failure modes of the selected equipment item
- Determine the effects of each failure
- Determine the criticality of that failure
- Identify the indications that the failure has occurred
- Estimate the rates (either as a function of time or per mission) for that failure mode
- Identify the failure compensation mechanisms

Table 4.4 FMEA Keywords.

Rupture	Crack
Leak	Plugged
Failure to open	Failure to close
Failure to stop	Failure to start
Failure to continue	Spurious stop
Spurious start	Loss of function
High pressure	Low pressure
High temperature	Low temperature
Overfilling	Hose bypass
Instrument bypassed	

The causes of equipment failures are not failure modes *per se*. For example, fouling of the tubes of a heat exchanger is not a failure, but it leads to the failure mode of insufficient cooling.

The guidewords in Table 4.4 provide suggestions for general failure modes.

The consequences of failures need to be developed in as much detail as possible. For example, if the failure mode is, "Pump fails to run," there is a world of difference between the pump tripping out for a few moments and a complete loss of pumping capability due to massive corrosion of the pump's impeller.

In general, the FMEA method does not consider the following issues:

- The reason for the occurrence of a particular failure mode
- Time-dependency and dynamic conditions
- Human error
- A sequence of events

Use of 3D models

Traditionally, the hazards analysis team would use P & IDs and PFDs for their base documentation. However, given that a large fraction of the hazards on an offshore facility are about layout and the movement of equipment and people, it has been found that it is more useful to project a 3D model of the rig or platform on an overhead screen so that people can visualize what is being talked about.

Job safety analysis

BOEMRE states the following about JSAs:

> *A JSA is used to review detailed job steps and uncover hazards associated with the specific job undertaken. The JSA defines the requirements for identifying, assessing, and controlling personal risks associated with work activities.*

Operators must complete a JSA prior to performing any activity identified in their SEMS program. The supervisor of the person in charge of the task must approve the JSA prior to the work commencing. The JSA is performed to identify and evaluate hazards of a job/task for the purpose of hazards control or elimination that is currently not addressed in API RP 75, Section 3, Hazards Analysis element.

Basically, a JSA is simply a hazards analysis that is directed toward tasks involving small numbers of people, or people working by themselves. The JSA identifies the hazards, determines the level of risk associated with those hazards, and requires the development of procedures to reduce risk to acceptable levels. JSAs are dynamic, i.e. they should be modified and adjusted as working conditions change.

JSAs are often one of the first steps taken in the development of operating procedures.

An early publication in this area (United States Steel, 1968) divided a JSA into the following four steps:

(1) Break the job or task down into steps or elements

(2) Determine the contact possibilities

(3) Eliminate or reduce contact possibilities

(4) Develop safe job procedures

In their *Accident Prevention Manual for Industrial Operations* the National Safety Council (NSC, 2009) provides a form that has the following three steps:

(1) Sequence of Basic Job Steps

(2) Potential Hazards

(3) Recommended Actions or Procedures

(It could be argued that the word "potential" in the second step is redundant. Hazards exist—they have the *potential* to cause an incident.) Leaving aside this semantic point, the NSC lists the following likely hazards:

- Could a person be struck by an object?
- Could a person be caught between two objects?
- Is there a potential for tripping, slipping, or falling?
- Could a person be injured due to activities such as pushing, pulling, lifting, bending, or twisting?
- Is the environment free of hazards such as toxic gases, mists, fumes, dust, heat, or radiation?

JSAs are typically developed by a supervisor, with input from those doing the work. For a JSA to be effective, Glenn (2011) notes that those preparing the JSA should have some knowledge of ergonomics so that they can determine whether a task could cause excessive strain. Also, those responsible for the JSA should have sufficient authority to approve changes to equipment layout, the availability of tools, and other issues so that they can reduce risk to an acceptable level.

Hazard register

The use of PSM systems, including SEMP and SEMS, results in the generation of identified hazards—usually in considerable quantity. They can come from hazards analyses, management of change, incident investigations, and audits. A system for tracking the findings and making sure that they are properly closed out is required. Such a system, as a minimum, should contain the following:

- A description of the hazard.
- Identification as to when and where the hazard was identified.
- Its risk ranking (not always necessary).
- Proposed action item to either eliminate the hazard, or to reduce its risk to an acceptable level (there may be more than one response).
- Who is responsible for closing out the item.
- The date by which closure is required.
- What was actually done, and when it was done.

For each identified hazard a full description is provided. The register contains a section for follow-up, as shown in Table 4.5.

Table 4.5 Sample Hazard Register.

Finding	Notes
Finding number	
Node	
Date of finding	
Hazard	
Source	
Consequences	
Safety	
Environmental	
Health	
Economic	
Likelihood	
Risk Rank	
Follow-up	
Assigned to	
Company	
Department	

Table 4.5 Sample Hazard Register—*Cont'd*

Finding	Notes
Recommendation	
Status	
Resolution	
Date approved	
Approved by	

The follow-up section describes how the identified hazard was handled, and when the associated recommendation was completed. On a large project it is necessary to have one person who is assigned the task of making sure that all findings are closed out properly before the new facility is started up. In addition to managing the risk register itself, the person in charge of follow-up generally is assigned the broader responsibility of filing all of the hazards analysis reports. Questions that have to be answered in this context include:

- How are the hazards analysis records to be managed?
- How are the recommendations and action items to be managed?
- How are the recommendations to be communicated?
- What media are to be used for storing the hazards analysis records?
- How and when are they to be purged?
- Who has access to the hazards analysis records?
- Who can modify the hazards analysis records?

MANAGEMENT OF CHANGE

Management of change (MOC) lies at the heart of any successful risk management program. Leaving aside malicious acts such as sabotage and terrorism, it can be taken for granted that everyone associated with the design and operation of an industrial facility wants to do a good job—yet, in spite of their best intentions, accidents continue to happen: people get hurt, production is lost, and the environment is polluted. Virtually all of these undesired events are caused by uncontrolled change. Someone, somewhere moved operating conditions outside their safe range without taking proper precautions.

Many of the comments that are made in this section with regard to MOC revolve around the definition of the word "change" and the related concepts of "in kind" and "not in kind" changes. This definition is more tricky than it may sound. When pushed to the limit all changes can be seen as not being "in kind." For example, two "identical" equipment item parts will have a different history—they could have been made at different times, or in different factories by different people.

MOC is often considered to be one of the most difficult of the elements of process safety to implement and control. For example, at the 7th Global Congress for Process Safety held in Chicago in March 2011 one of the sessions was entitled *Management of Change – The Most Difficult PSM Challenge.* After 20 years of PSM on onshore process plants, these industry professionals were saying that MOC is still extremely challenging.

Regarding cost, a rule of thumb for onshore facilities is that each change processed through the MOC system has a direct administrative cost of around $1500, with associated implementation costs of at least $1 million for a 400-person facility.

The SEMS rule

SEMS Standard	Discussion
3.1 General	
Management of Change (MOC) procedures are needed to ensure that any changes to the facility are analyzed for hazards, and that safety information is updated. Particular attention needs to be paid to temporary changes. Management and facility personnel that even the most minor of changes could have serious consequences if not properly managed.	The guidance provided in this section rightly notes that facilities are subject to continual change. For this reason it is vital that the MOC program be implemented at the very beginning of a SEMP or SEMS program.
3.2 Change in Facilities	
This section starts by saying, "Change in facilities arises whenever the process or mechanical design is altered." The section then goes on to list the many types of changes that can occur. These include changes to: • Products and produced fluids; • Product specifications; • By-products and waste products; • Construction projects; • Mechanical and equipment changes; • Changes to operating conditions; • Bypasses; • The use of new chemicals; and • Modifications to drilling systems.	This section lists many of the items that should trigger an MOC request. These items can be compiled into the form of a checklist. Sub-paragraph e. refers to "significant changes in operating conditions, including pressures, temperatures, flow rates, or process conditions". In the context of MOC, the word "significant" is self-referential. In order to determine if a proposed change is significant enough to require the application of MOC, it is important to provide criteria for "Change" as distinct from "change". This important and difficult topic is discussed in Chapter 13 of *Process Risk and Reliability Management.* With regard to minor changes, some judgment is required. Not all such changes need go through the MOC process. The reference to drilling systems is needed because the concepts behind Management of Change were developed for

SEMS Standard	Discussion
	continuously-operating units. Drilling activities involve equipment and practices that do not neatly fit inside the production envelope.
3.3 Change in Personnel	
MOC is required if there is a change in the organization (including contractors) that could affect the facility's safety. Routine personnel changes are handled through other elements of SEMS, including operating procedures, safe work practices, and training.	Based on experience with onshore facilities, MOC is frequently applied to changes in the organizational structure, but not to changes in specific individuals. MOC would also be applied if the number of people working in a particular area is to be changed, or if there are significant changes in operating conditions (for example, a move from eight-hour to twelve-hour shifts).
The standard notes that significant organizational changes, including change of ownership, may generate the need for an MOC.	It is common for an acquiring company to conduct a detailed assessment of potential environmental liabilities as part of the due diligence process. Increasingly, a SEMP/SEMS evaluation will be part of that process also.
3.4 Managing the Changes	
There need to be formal procedures for evaluating and managing changes. The procedures should consider changes to the equipment design, safety and health impacts, communications with personnel (including operating procedures and training), the duration of temporary changes, and authorizations.	The key to this paragraph is that the Management of Change system must be written down. Basically, this section calls on companies to develop a checklist of items to consider when reviewing a potential change. The need for a hazards analysis does imply some circular logic. Only after a hazards analysis has been completed does management know if the analysis was actually needed. In practice, the judgment and experience of experienced personnel will have to be relied upon. In practice, this is often one of the areas where MOC programs tend to fall short. It is all too easy to put off work on procedures and training "until we have enough time". It is very important to manage temporary changes properly. There can be a tendency to say, "This change will last for only an hour or two, so we don't need to examine it as thoroughly as if it were permanent." Also, temporary changes have a tendency to become permanent. For example, it may be decided to put a "temporary" bypass around a control valve so that the valve can be pulled for maintenance. Once the valve is back in service the bypass may be left in place.

BOEMRE requirements

§250.1912 What criteria for management of change must my SEMS program meet?

BOEMRE Requirements	Discussion
(a) You must develop and implement written MOC procedures for modifications associated with the following: (1) Equipment, (2) Operating procedures, (3) Personnel changes (including contractors), (4) Materials, and (5) Operating conditions.	What is written here aligns with the SEMP requirements discussed above.
(b) MOC procedures do not apply to situations involving replacement in kind (such as, replacement of one component by another component with the same performance capabilities).	Determining when a replacement is "in kind," and when it is not is tricky. For example, a facility may replace a gasket with another that is identical, and so "in kind." However, the new gasket was probably made at a different time, possibly in a different factory, by different people, and was shipped and stored differently. Any of these changes could be significant enough to cause a failure of that gasket when it is put in service.
(c) You must review all changes prior to their implementation.	This requirement is handled through the prestart-up safety review/operational readiness programs.
(d) The following items must be included in your MOC procedures: (1) The technical basis for the change; (2) Impact of the change on safety, health, and the coastal and marine environments; (3) Necessary time period to implement the change; and (4) Management approval procedures for the change.	Once more, these requirements are a repeat of what is written in the SEMP standard.
(e) Employees, including contractors whose job tasks will be affected by a change in the operation, must be informed of, and trained in, the change prior to start-up of the process or affected part of the operation;	
(f) If a MOC results in a change in the operating procedures of your SEMS program, such changes must be documented and dated.	Many changes will result in a change to operating procedures. Even a "pure" equipment change, such as increasing the size of a pump impeller, will result in different operating conditions. Hence, updated procedures (and training) will be required.

Management of change offshore

The MOC offshore does not have any special characteristics, although the dynamic environment associated with a drilling program can make it difficult to be sure when a change should be treated as a "Change" under the MOC program. For example, the mud weight used during a drilling program will vary according to the depth of the hole. Yet BSEE states, the MOC must be applied for any change in a facility, and such a change includes:

Introduction of new or different[...] drilling muds[...]

In practice, if the drilling program calls for a range of mud weights and compositions to be used as the well gets deeper, then only a change outside that range would constitute a "Change."

OPERATING PROCEDURES

No matter how automated processes become, there will always be a man—machine interface, and that interface has to be controlled through the use of operating procedures and training.

The SEMP standard discusses the use of standardized operating procedures. Such procedures that apply, with only minor variations, to facilities that differ only in size or the number of equipment items present. The standard also requires that human factors issues (page 233) to do with format, content, and intended use should be incorporated into the procedures.

Detailed guidance to do with the development and implementation of operating procedures is provided in Chapter 9 of *Process Risk and Reliability Management.*

The SEMS rule

The SEMS standard to do with operating procedures is described and discussed below. It cites the following documents:

(1) API RP 14 J, Recommended Practice for Design and Hazards Analysis for Offshore Production Facilities (latest edition), for information on start-up, normal operations, and shutdown of production facilities.

(2) MODU Operations Manual developed in conformance with flag state requirements and/or the IMO MODU Code for information on routine operations and operating limits on MODUs.

SEMS Standard	Discussion
4.1 General	
The standard requires that operating procedures are necessary for all operations. Standard, or generic procedures, can be used for facilities that are similar to one another. Once more, there is a reference to human factors.	Many companies use the term Standard Operating Procedures (SOPs) for all of their operating procedures. This paragraph (which is matched by something similar in the EPA Risk Management Program standard) draws a distinction between *standard* operating

SEMS Standard	Discussion
	procedures and *task-specific* operating procedures. A standard procedure will describe a general activity, such as starting a pump or lifting a heavy object over the deck. A task-specific procedure describes the actions needed to start a particular pump, or to lift a particular heavy object.
4.2 Content of Operating Procedures	
The procedures should be consistent with the Safety and Environmental Information, and should cover all phases of the operation, including temporary and simultaneous operations. References to other API standards are provided.	The requirement for describing a person's "reporting relationship" is an unusual aspect of operating procedures.
The procedures should be written such that operations stay within the operating limits as provided in the Safety and Environmental Information section. The procedures should also provide guidance on the prevention of environmental damage, and what to do if there is a release.	The requirements of this section are both important and difficult. It is often difficult to know what the numerical limits on the safe operating envelope actually are.
4.3 Periodic Review	
Operating procedures should always be reviewed as part of the MOC process. There should also be a regular review of the procedures (the frequency depending on the hazards of the process covered).	The standard does not specify a specific time period for routine procedure review. It suggests that one of the roles of the hazards analysis team is to help define this schedule. The communication of changes to procedures will generally be incorporated into the training program.

BOEMRE requirements

§250.1913 What criteria for operating procedures must my SEMS program meet?

BOEMRE Requirements	Discussion
(a) You must develop and implement written operating procedures that provide instructions for conducting safe and environmentally sound activities involved in each operation addressed in your SEMS program. These procedures must include the job title and reporting relationship of the person or persons responsible for each of the facility's operating areas and address the following: (1) Initial start-up;	The 12 paragraph elements listed here can form the basis for the table of contents for the operating procedures manual.

BOEMRE Requirements	Discussion
(2) Normal operations; (3) All emergency operations (including but not limited to medical evacuations, weather-related evacuations, and emergency shutdown operations); (4) Normal shutdown; (5) Start-up following a turnaround, or after an emergency shutdown; (6) Bypassing and flagging out-of service equipment; (7) Safety and environmental consequences of deviating from your equipment operating limits and steps required to correct or avoid this deviation; (8) Properties of, and hazards presented by, the chemicals used in the operations; (9) Precautions you will take to prevent the exposure of chemicals used in your operations to personnel and the environment. The precautions must include control technology, personal protective equipment, and measures to be taken if physical contact or airborne exposure occurs; (10) Raw materials used in your operations and the quality control procedures you used in purchasing these raw materials; (11) Control of hazardous chemical inventory; and (12) Impacts to the human and marine environment identified through your hazards analysis.	
(b) Operating procedures must be accessible to all employees involved in the operations.	Generally, procedures will be supplied electronically. An operator can print any particular procedure for use outside. However, the printed copy should have the word "Uncontrolled" on it, so that it will not be used at a later time when it may have been replaced by a newer version.
(c) Operating procedures must be reviewed at the conclusion of specified periods and as often as necessary to assure they reflect current and actual operating practices, including any changes made to your operations.	OSHA requires that operating procedures be revalidated every year. This requirement does not apply offshore, but it provides guidance. In practice, procedures are often updated on an ongoing basis because most of the changes required by the MOC process require

BOEMRE Requirements	Discussion
	that one or more of the operating procedures be updated.
(d) You must develop and implement safe and environmentally sound work practices for identified hazards during operations and the degree of hazard presented.	This requirement is covered by the need to develop JSAs and the safe work practices discussed in the next section.
(e) Review of and changes to the procedures must be documented and communicated to responsible personnel.	This is a repeat of the same statement to be found in SEMP.

SAFE WORK PRACTICES

Safe work practices cover operating, maintenance, and material handling risks. Generally, work in this area would fall under the category of occupational safety. Many of the best practices are described in API 76, *Improving Owner and Contractor Safety Performance.* Human factors can be included in this topic.

The SEMS rule

The SEMS requirements to do with safe work practices are described and discussed below.

SEMS Rule	Discussion
5.1 General	
Safe works practices are required. They should cover all operating and maintenance work, particularly those that could affect safety or have an environmental impact.	Generally, these safe work practices will be standard procedures, analogous to the development of Standard Operating Procedures (SOPs) discussed above. Generic safe work practices are needed for activities such as: • Opening of pressurized or energized equipment or piping; • Lockout and tagout of electrical and mechanical energy sources; • Hot work; • Confined space entry; and • Crane operations.
Human factors are mentioned once again. Reference is made to API RP 76 *Improving Owner and Contractor Safety Performance.*	As already discussed, neither SEMP nor SEMS provide much specific guidance to with the topic of human factors.
The standard states that, "Contractors should have their own written safe work practices.	It is important that there is proper coordination between the safe work practices of the operator

SEMS Rule	Discussion
Contractors may adopt appropriate sections of the operator's safety and environmental management program."	and of the contractors working at the facility. This is an important part of the mapping process. Some training companies provide training in generic safe work practices such as lockout/tagout or vessel entry. This training is then accepted by a wide range of operating companies, thus minimizing the training investment required.
5.2 Safe Conduct of Work Activities	
Safe work practices will involve the use of a work authorization or permit to work system. The system should describe how work activities are communicated at shift change, and when new personnel are involved.	Although crane operations are omitted from the list of required permits to work, most companies do require that a plan and permit be prepared for all lifting operations, particularly if equipment is being lifted over operating equipment or areas where hazardous or flammable chemicals are stored.
5.3 Control of Hazardous Materials	
Information about hazardous materials should be included in the safe work practices.	This information will generally be found in the facility's MSDS (Material Safety Data Sheets), which are part of the Safety and Environmental Information.
5.4 Contractor Selection	
Information regarding a contractor's safety and environmental policies and performance should be included in the contractor selection decision process. The operator's own safety and environmental policies should be communicated to contractors. API RP 76 is a useful reference for selecting contractors.	The manner in which contractors were selected and evaluated will have to be properly documented so that the process can be audited. As has already been noted, the contractor could be a very large company, maybe much larger than the operator to which it is providing services. In such cases, the operator may find it necessary to adapt its procedures to those of the contractor.

BOEMRE requirements

§250.1914 What criteria must be documented in my SEMS program for safe work practices and contractor selection?

BOEMRE Requirements	Discussion
Your SEMS program must establish and implement safe work practices designed to minimize the risks associated with operating, maintenance, and modification activities and the	This paragraph is largely a repeat of what is provided in SEMP.

BOEMRE Requirements	Discussion
handling of materials and substances that could affect safety or the environment. Your SEMS program must also document contractor selection criteria. When selecting a contractor, you must obtain and evaluate information regarding the contractor's safety and environmental performance.	
Operators must ensure that contractors have their own written safe work practices. Contractors may adopt appropriate sections of the operator's SEMS program. Operator and contractor must document their agreement on appropriate contractor safety and environmental policies and practices before the contractor begins work at the operator's facilities.	The process for bridging the safe work practices of the operator and contractor is important. It is discussed on page 119.
(a) A contractor is anyone performing work for the lessee. However, these requirements do not apply to contractors providing domestic services to the lessee or other contractors. Domestic services include janitorial work, food and beverage service, laundry service, housekeeping, and similar activities.	
(b) You must document that your contracted employees are knowledgeable and experienced in the work practices necessary to perform their job in a safe and environmentally sound manner. Documentation of each contracted employee's expertise to perform his/her job and a copy of the contractor's safety policies and procedures must be made available to the operator and BOEMRE upon request.	This is a challenging requirement, and needs to be integrated into the bridging process.
(c) Your SEMS program must include procedures and verification for selecting a contractor as follows: (1) Your SEMS program must have procedures that verify that contractors are conducting their activities in accordance with your SEMS program. (2) You are responsible for making certain t hat contractors have the skills and knowledge to perform their assigned duties and are conducting these activities in accordance with the requirements in your SEMS program.	

BOEMRE Requirements	Discussion
(3) You must make the results of your verification for selecting contractors available to BOEMRE upon request.	
(d) Your SEMS program must include procedures and verification that contractor personnel understand and can perform their assigned duties for activities such as, but not limited to: (1) Installation, maintenance, or repair of equipment; (2) construction, start-up, and operation of your facilities; (3) turnaround operations; (4) major renovation; or (5) specialty work.	This requirement is discussed in the SEMP section.
(e) You must: (1) Perform periodic evaluations of the performance of contract employees that verifies they are fulfilling their obligations, and (2) Maintain a contractor employee injury and illness log for 2 years related to the contractor's work in the operation area, and include this information on Form MMS-131.	In practice, this requirement is challenging. An operator may hire a contractor company that it has never used before. Therefore, a 2 year's log of injury and illness will not necessarily be available.
(f) You must inform your contractors of any known hazards at the facility they are working on including, but not limited to fires, explosions, slips, trips, falls, other injuries, and hazards associated with lifting operations.	
(g) You must develop and implement safe work practices to control the presence, entrance, and exit of contract employees in operation areas.	

TRAINING

All aspects of operational integrity and process safety come down to people working with systems; and people—including contract workers—need to be trained in running those systems. Training is expensive and time-consuming, but, in the words of Zig Ziglar, "The only thing worse than training your employees and losing them is NOT training your employees and keeping them."

SEMS does not use the word "competence," but it does refer to skills and knowledge, which would seem to amount to much the same thing.

The SEMS rule

The SEMS requirements to do with training are described and discussed below.

SEMS Rule	Discussion
6.1 General	
Training that is provided by properly qualified instructors should be provided to all personnel who are working offshore. The training program should be fully integrated with the Management of Change program.	This paragraph identifies the strong link between procedures and training. They are basically two sides of the same coin. There is little point in having procedures if the affected employees are not trained in their use. Equally, it is impossible to provide equipment-specific training without having good quality procedures. Although this paragraph does not use the term 'Management of Change', it is what it is all about.
6.2 Initial Training	
6.2.1 Basic Training	
Basic training includes the orientation provided in API RP T-1, *Recommended Practice for Orientation Program for Personnel Going Offshore for the First Time,* API RP T-4, *Recommended Practice for Training of Offshore Personnel in Non-Operating Emergencies,* API RP T-7, *Recommended Practice for Training of Personnel in Rescue of Persons in Water* and API RP 14 G, *Recommended Practice for Fire Prevention and Control on Open Type Offshore Production Platforms.* The training should include non-operating emergencies and safe work practices such as hot work and lockout/tagout. The training should also cover the pertinent government regulations.	
6.2.2 Qualification Criteria	
Training should include the development and application of qualification criteria.	In addition to generic training, personnel need to be trained in the facility's operating procedures: both standard and task-specific. The standard refers to many of the well-established API documents to do with training.
The initial training should include an introduction to environmental protection and pollution control.	For most personnel, this requirement involves little more than making sure that everyone knows that dropping objects overboard is completely unacceptable.

SEMS Rule	Discussion
	However, senior operators and supervisors need to be aware of the allowable limits for discharges of materials such as produced water, and they must know when they seem to be not in compliance with those limits.
6.3 Periodic Training/ 6.4 Communication	
Refresher training is required, particularly when procedures or equipment change.	This requirement should be self-evident.
6.5 Contractor Training	
Contractors are required to train their personnel.	Although contractors are responsible for training their people, the ultimate SEMS responsibility for safe operations rests with the facility operator.
Contractors who provide incidental services that do not influence operation of the facility require basic training in basic safety and emergency response procedures.	It is noteworthy that many of the men who died in the Piper Alpha catastrophe were trapped in the Living Quarters, and had not received the basic safety training described here.

BOEMRE requirements

§250.1915 What criteria for training must be in my SEMS program?

BOEMRE Requirement	Discussion
Training must address the operating procedures (§250.1913), the safe work practices (§250.1914), and the emergency response and control measures (§250.1918). You must document the qualifications of your instructors. Your SEMS program must address:	
(a) Initial training for the basic well-being of personnel and protection of the environment, and ensure that persons assigned to operate and maintain the facility possess the required knowledge and skills to carry out their duties and responsibilities, including start-up and shutdown.	
(b) Periodic training to maintain understanding of, and adherence to, the current operating procedures, using periodic drills, to verify adequate retention of the required knowledge and skills.	

BOEMRE Requirement	Discussion
(c) Communication requirements to ensure that whenever a change is made to operating procedures (§250.1913), the safe work practices (§250.1914), or the emergency response and control measures (§250.1918), personnel will be trained in or otherwise informed of the change before they are expected to operate the facility.	
(d) How you will verify that the contractors are trained in the work practices necessary to perform their jobs in a safe and environmentally sound manner, including training on operating procedures (§250.1913), the safe work practices (§250.1914), or the emergency response and control measures (§250.1918).	The verification of training programs can follow the guidance provided in the existing subpart "O" regulation that is discussed below.

Subpart "O"

In the year 2000, the MMS published 30 CFR Part 250, subpart O—Training. The regulation calls for each lessee to prepare a training plan laying out the company's training philosophy including the type, method, length, frequency, and content of their training program. To implement this regulation BOEMRE has used a series of performance measures to periodically assess the quality of lessee and contractor training programs. Such assessments have included one or more of the following evaluation methods:

- Audits
- Written tests
- Hands-on tests
- Employee interviews

In the Introduction to the document the MMS stated:

Subpart O is MMS' first truly performance-based regulation. It could have wide implications for the future of our regulatory program. The new rule eliminates the need for school certification; it is the lessee's responsibility to develop and implement their training program according to broad guidelines presented in the rule and the specific needs of their company. It is less prescriptive than the old rule in that it does not set out specific requirements for the operator to follow in regards to the frequency, length, topics or methods of their training.

The new rule also addresses contractors through the lessee-training plan.

The following explanation is also provided by the MMS:

The goal of your training program must be safe and clean OCS operations. To accomplish this, you must ensure that your employees and contract personnel engaged in well control or production safety operations understand and can properly perform their duties.

Subpart O requires that you evaluate the Subpart O training programs of your contractors and that you verify that all contractor personnel engaged in well control or production safety operations can perform their assigned duties. In addition, upon request of the MMS, you must be able to provide copies of training documentation for personnel involved in well control or production safety operations for a 5-year period.

With regard to SEMS, subpart O requirements are still in effect, but they are considered as being one component of the overall SEMS-required training.

SafeGulf

The SafeGulf program is endorsed by some of the leading operators in the Gulf of Mexico. Under this program, all contractors must fulfill minimum HSE training requirements prior to working on operator premises in the Gulf of Mexico. The program, which is provided by various companies and organizations, provides minimum training and orientation. A database is maintained of all those who have had this training.

The SafeGulf basic training is supplemented by specialized training, as needed. Some of this specialized training takes the form of industry certification programs.

As part of the response to the SEMS rule, companies operating in the Gulf of Mexico are considering expanding the scope of the SafeGulf program.

MECHANICAL INTEGRITY

This topic—whose full name is "Assurance of Quality and Mechanical Integrity of Critical Equipment"—requires that procedures are in place and implemented so that critical equipment is designed, fabricated, installed, tested, inspected, monitored, and maintained in a manner consistent with appropriate service requirements, manufacturer's recommendations, or industry standards. Contractors should have programs in place to address the safety of their own critical equipment.

As noted above in the discussion regarding hazards analysis, offshore facilities do not generally handle highly corrosive or toxic chemicals, hence reducing the associated integrity problems. However, offshore platforms operate in a marine environment, where much of the equipment is exposed to salt water and harsh environmental conditions.

Human factors should be considered, particularly regarding equipment accessibility for operation, maintenance, and testing. The overall quality assurance strategy to require conformance to specifications/requirements should be developed at the beginning of the project and become a part of the overall project execution plan and maintenance program. The quality assurance strategy should carry over into the operating and maintenance procedures and MOC.

The SEMS rule

The SEMS requirements to do with mechanical integrity are described and discussed below.

SEMS Rule	Discussion
7.1 General	
All critical equipment should be subject to a full integrity program. Human factors and quality assurance should both be considered.	The mechanical integrity of equipment is obviously a crucial element in any SEMS program. The ultimate purpose of such programs is to ensure that hazardous and flammable materials remain confined with the system's piping and vessels. The mechanical integrity program, therefore, represents the last line of defense. BOEMRE does not define the meanings of the words 'critical' or 'significant' when referring to equipment. They leave that to the operator (Levine, 2011). However, a 'critical equipment' item would generally be one that must be working for the facility to be in operation. From a practical point of view, the implementation of a Mechanical Integrity program requires careful coordination with on-going maintenance and inspection activities. Also, it can be assumed that this term will cover all safety devices (both mechanical and instruments).
7.2 Procurement	
The mechanical integrity program covers the procurement of equipment.	A paper trail needs to be in place to show that the equipment that was actually purchased meets the original design and material specifications. This paper trail can be verified as part of the Pre-Startup Review process.
7.3 Fabrication/7.4 Installation	
Quality control procedures for fabrication and installation are required.	The comments above to do with procurement (have a paper trail and integrate with the Pre-Startup Review) appear to apply equally well to the fabrication phase of a project.
7.5 Maintenance	
Maintenance is part of the Mechanical Integrity program. It covers the personnel, both operator and contractor, involved in the maintenance program, including their training and procedures.	

SEMS Rule	Discussion
7.6 Testing and Inspection	
Critical equipment needs to be tested on a regular basis. The standard cites API 510, *Pressure Vessel Inspection Code: Maintenance Inspection, Rating, Repair, and Alteration.* Programs may be required for environmental protection compliance monitoring.	This requirement should be self-evident. Once more, the 'critical equipment' issue comes up. One way of developing such a list may be through the hazards analysis process. The team can be charged with explicitly identifying which equipment items are deemed to be critical.

BOEMRE requirements

§250.1916 What criteria for mechanical integrity must my SEMS program meet?

BOEMRE Requirement	Discussion
You must develop and implement written procedures that provide instructions to ensure the mechanical integrity and safe operation of equipment through inspection, testing, and quality assurance. The purpose of mechanical integrity is to ensure that equipment is fit for service. Your mechanical integrity program must encompass all equipment and systems used to prevent or mitigate uncontrolled releases of hydrocarbons, toxic substances, or other materials that may cause environmental or safety consequences. These procedures must address the following:	The requirements in this section are for thorough documentation. This will minimize the chance of oversights or slippages, and will allow an auditor to assess what is going on.
(a) The design, procurement, fabrication, installation, calibration, and maintenance of your equipment and systems in accordance with the manufacturer's design and material specifications.	The facility's procedures need to coordinate with the manufacturer's documentation and instructions.
(b) The training of each employee involved in maintaining your equipment and systems so that your employees can implement your mechanical integrity program.	
(c) The frequency of inspections and tests of your equipment and systems. The frequency of inspections and tests must be in accordance with BOEMRE regulations and meet the manufacturer's recommendations. Inspections and tests can be performed more frequently if determined to be necessary by prior operating experience.	
(d) The documentation of each inspection and test that has been performed on your equipment and systems. This documentation must identify the date of the inspection or test; include the name and position, and the signature of the person who performed the inspection or test; include the serial number or other identifier of the equipment on which the inspection or test was performed; include a description of the inspection or test performed; and the results of the inspection test.	

BOEMRE Requirement	Discussion
(e) The correction of deficiencies associated with equipment and systems that are outside the manufacturer's recommended limits. Such corrections must be made before further use of the equipment and system.	
(f) The installation of new equipment and constructing systems. The procedures must address the application for which they will be used.	
(g) The modification of existing equipment and systems. The procedures must ensure that they are modified for the application for which they will be used.	
(h) The verification that inspections and tests are being performed. The procedures must be appropriate to ensure that equipment and systems are installed consistent with design specifications and the manufacturer's instructions.	
(i) The assurance that maintenance materials, spare parts, and equipment are suitable for the applications for which they will be used.	

PRESTART-UP REVIEW

The purpose of a prestart-up or operational readiness review is to ensure that initial start-ups, or start-ups following major project work, proceed safely and smoothly. The review team will make sure that management systems are in place, and have been followed, so that the plant will operate safely on restart.

The review has two major components:

(1) Ensure that all action items and recommendations from hazards analyses and other reviews (such as management of change) have been completed as required.

(2) Ensure that no changes that could affect safety or operability have crept into the system during the construction phase.

The review represents the last chance to catch any problems. Therefore, it should be carried out by the operations personnel who will be required to run the modified system.

Generally, the following issues will be covered by the review team:

- Equipment and instrumentation that have been changed were installed and commissioned in accordance with design specifications.
- Safety, operating, maintenance, and emergency procedures are in place and are adequate.
- All findings from hazards analyses, MOC evaluations, and other types of review have been closed out properly.
- All affected personnel have been trained in the new or modified operation.

The SEMS rule

The SEMS requirements for prestart-up safety reviews are described and discussed below. Guidance to do with prestart-up reviews is provided in API RP 75, Recommended Practice for Design and Hazards Analysis for Offshore Production Facilities.

SEMS Rule	Discussion
8.1 General	
This section of SEMP discusses the requirements of a pre-startup safety review. It refers to APPI RP 14 J and to Appendix B of the standard.	The purpose of a pre-startup safety or operational readiness review is to ensure that initial start-ups, or start-ups following major project work, proceed safely and smoothly. The review team will make sure that management systems are in place, and have been followed, such that the plant will operate safely on restart. The review has two major components: 1. Ensure that all action items and recommendations from Hazards Analyses and other reviews (such as Management of Change) have been completed as required. 2. Ensure that no changes that could affect safety or operability have crept into the system during the construction phase. The review represents the last chance to catch any problems. Therefore it should be carried out by the operations personnel who will be required to run the modified system. Generally, the following issues will be covered by the review team: • Equipment and instrumentation that have been changed were installed and commissioned in accordance with design specifications. • Safety, operating, maintenance, and emergency procedures are in place and are adequate. • All findings from hazards analyses, management of change evaluations and other types of review have been closed out properly. • All affected personnel have been trained in the new or modified operation.
	Readiness reviews are important because projects frequently fall behind schedule and/or run over budget, thus creating pressure on the project team to eliminate or postpone the installation of any items that are not absolutely necessary for the start-up. If not controlled properly, this can lead to corner-cutting — either intentional or inadvertent — which may in turn jeopardize the safety or operability of the modified facility. The review gives the operations department the authority to refuse to accept 'care, custody and control' of a facility that they judge to be unsafe or difficult to operate.

SEMS Rule	Discussion
	It is not the purpose of a Pre-startup Review to replace this normal 'punching out' of the facility. Pre-startup reviews cover not only equipment, but also 'soft' issues, such as operating procedures and training. These are particularly important with regard to restart reviews – it is critical that the operating procedures are updated to reflect the changes that have been made, and that the operators are then trained in the new procedures before they start work on the modified facility. Pre-startup reviews sometimes identify documentation problems. During the construction and commissioning of a plant there is usually considerable pressure just to get the plant completed and up and running. Hence the record-keeping part of the project may slip. If, in the judgement of the review team, some of the missing documents are important to safety, they must ensure that those documents are completed and issued before the plant is started. A pre-startup review is not a last-minute hazards analysis. The review team merely checks that a professional-level hazards analysis was carried out on the changed facility, and that all the findings were implemented or closed out in a professional manner. The team does not actually analyze the new system for hazards.

BOEMRE requirements

§250.1917 What criteria for prestart-up review must be in my SEMS program?

BOEMRE Requirement	Discussion
Your SEMS program must require that the commissioning process include a prestart-up safety and environmental review for new and significantly modified facilities that are subject to this subpart to confirm that the following criteria are met:	This section is essentially a repeat of the SEMP material.
(a) Construction and equipment are in accordance with applicable specifications.	
(b) Safety, environmental, operating, maintenance, and emergency procedures are in place and are adequate.	
(c) Safety and environmental information is current.	
(d) Hazards analysis recommendations have been implemented as appropriate.	

BOEMRE Requirement	Discussion
(e) Training of operating personnel has been completed.	
(f) Programs to address MOC and other elements of this subpart are in place.	
(g) Safe work practices are in place.	

EMERGENCY RESPONSE AND CONTROL

The purpose of any type of PSM program is to make sure that emergencies do not occur. However, emergencies do occur, even if only rarely, so an emergency response system is needed.

The amount of space on most offshore platforms is extremely limited. This makes emergency response difficult because access to affected equipment items can be very restricted. In addition, there is nowhere for personnel on the platform to escape to; unlike those working in an onshore refinery or chemical plant who can evacuate the affected areas. Such an evacuation is much more difficult offshore, and usually involves the use of lifeboats, life-rafts, and escape devices that can themselves pose significant dangers.

The facility should have a written action plan that assigns authority to a designated person who will initiate and manage the emergency response and follow-up reporting. The plan should consider issues such as the following:

- Spills of hazardous substances
- Collisions
- Fires and explosions
- Blowouts
- Contingency plans

It is vital that the emergency plans be tested on a regular basis.

The SEMS rule

The SEMS requirements concerning emergency response and control are described and discussed below.

SEMS Rule	Discussion
9.1 General	
This section discusses the need for an emergency response and control plan and the importance of regular drills.	A written plan is of no value unless personnel on a facility know what to do as soon as there is an emergency. Therefore, it is very important that

SEMS Rule	Discussion
	drills are carried out frequently. Many incidents – most notably Piper Alpha – were made much worse because the people involved did not know what to do following the initial explosion. However, the drills should not be too realistic. Fatal incidents have occurred when platform personnel were practising the launch and recovery of escape craft.
9.2 Emergency Action Plan	
This section describes the development of a plan and who is in charge during an emergency.	A clear line of command should be established for emergency operations. The line of command may differ from the normal management structure, although normally the OIM (Offshore Installation Manager) will lead the emergency response. The plan must contain a provision to make sure that all emergency situations are reported to the appropriate authorities in a timely and proper manner.
9.3 Emergency Control Center	
The role of the Emergency Control Center is described in this paragraph.	The Emergency Control Center (ECC) will often be located in the same general area as the Temporary Refuge. The Center should provide emergency response personnel with the capability of operating the control systems that are critical to the safety of the facility. Personnel working in the ECC should have access to all relevant plans. They should also be able to access safety and environmental information quickly.
9.4 Training and Drills	
The final section on Emergency Response and Control discusses training and drills.	It is critical that all personnel are trained in the emergency response procedures. Although, as discussed, drills that are too realistic can create serious safety problems, it is important not to have the drills become too routine and predictable. The training should consider the requirements of those crew members who are off duty, and who may be asleep in the Living Quarters.

BOEMRE requirements

§250.1918 What criteria for emergency response and control must be in my SEMS program?

BOEMRE Requirement	Discussion
Your SEMS program must require that emergency response and control plans are in place and are ready for immediate implementation. These plans must be validated by drills carried out in accordance with a schedule defined by the SEMS training program (§250.1915).	
The SEMS emergency response and control plans must include:	
(a) Emergency action plan that assigns authority and responsibility to the appropriate qualified person(s) at a facility for initiating effective emergency response and control, addressing emergency reporting and response requirements, and complying with all applicable governmental regulations;	
(b) ECC(s) designated for each facility with access to the Emergency action plans, oil spill contingency plan, and other safety and environmental information (§250.1910);	The ECC will often be located at the primary refuge or muster area. The center may not have full control room capability, but it must be provided with the equipment and instrumentation needed to bring the facility into a safe condition, to communicate with outside parties, and to coordinate the evacuation of personnel.
(c) Training and drills incorporating emergency response and evacuation procedures conducted periodically for all personnel (including contractor's personnel), as required by the SEMS training program (§250.1915). Drills must be based on realistic scenarios conducted periodically to exercise elements contained in the facility or area emergency action plan. An analysis and critique of each drill must be conducted to identify and correct weaknesses.	As already noted, drills can themselves be hazardous.

INVESTIGATION OF INCIDENTS

The thorough investigation and analysis of incidents (both actual events and near misses), along with the appropriate follow-up, provide one of the most effective means of improving the safety and reliability of process facilities. Other risk management programs, such as hazards analysis and MOC, are directed toward anticipating problems so that corrective actions can be taken before an event occurs. Actual incidents, on the other hand, provide hard information as to how things can go wrong, thus helping to cut through wishful thinking, prejudice, ignorance, and misunderstandings. The root cause analysis that follows an incident investigation will help identify weaknesses and limitations in a facility's management system, thereby reducing the chance of recurrence of similar incidents.

The SEMS rule

The SEMS requirements on the investigation of incidents are discussed below.

SEMS Rule	Discussion
10.1 General	
This section discusses the background to incident investigations, why they are needed and how corrective actions can be developed.	The most important aspect of incident investigation and analysis is that all incidents are properly evaluated; they must not be brushed aside as being part of 'business as usual'. The purpose of the investigation, which is generally a team effort, is to learn from what has happened so that it does not occur again. The analysis of the incident may lead to an evaluation of 'root causes'. As with hazards analysis, the key to successful incident investigation is having a team leader who has extensive industrial experience and who is properly trained in conducting investigations. Such people are hard to find.
10.2 Investigation	
The scope of a typical investigation is discussed.	Detailed guidance as to the organization and conduct of an incident investigation is discussed in Chapter 12 of *Process Risk and Reliability Management.*
10.3 Follow Up	
This section discusses the retention of incident investigation for use in subsequent hazards analyses, documentation of follow-up to findings, and the distribution of lessons learned to other, similar facilities.	Incident findings and recommendations can be kept in the same data base as is used for hazards analyses. After all, both activities are incident investigations – the only difference being that a hazards analysis examines what *might* happen, not what *has* happened. The distribution of learnings from incidents can raise some difficult legal issues, particularly when multiple companies are involved. Nevertheless, such learnings can be extremely valuable. One offshore contractor had one of its employees die in what was a totally unnecessary accident. Management of that company then sent those involved in the investigation to many other platforms in the same area to explain what had happened, and what lessons were to be learned.

BOEMRE requirements

§250.1919 What criteria for investigation of incidents must be in my SEMS program?

BOEMRE Requirement	Discussion
To learn from incidents and help prevent similar incidents, your SEMS program must establish procedures for investigation of all incidents with serious safety or environmental consequences and require investigation of incidents that are determined by facility management or BOEMRE to have possessed the potential for serious safety or environmental consequences. Incident investigations must be initiated as promptly as possible, with due regard for the necessity of securing the incident scene and protecting people and the environment. Incident investigations must be conducted by personnel knowledgeable in the process involved, investigation techniques, and other specialties that are relevant or necessary.	The content of this section is very much the same as that provided in the SEMP standard.
(a) The investigation of an incident must address the following: (1) The nature of the incident; (2) The factors (human or other) that contributed to the initiation of the incident and its escalation/control; and (3) Recommended changes identified as a result of the investigation.	
(b) A corrective action program must be established based on the findings of the investigation in order to analyze incidents for common root causes. The corrective action program must: (1) Retain the findings of investigations for use in the next hazard analysis update or audit; (2) Determine and document the response to each finding to ensure that corrective actions are completed; and (3) Implement a system whereby conclusions of investigations are distributed to similar facilities and appropriate personnel within their organization.	Although spreading the results of investigations to similar facilities can certainly help reduce the chance of additional incidents, consideration should be given to any liability concerns that may arise.

AUDITS

At the time of writing, various audit tool kits are being published. They are available through the International Association of Drilling Contractors and the Offshore Operators Committee. These documents help establish gap analyses for both operators and contractors.

All management programs, including those for managing offshore safety, must be audited on a regular basis, and a program for addressing the findings in a timely manner also needs to be in place. In the words of one facility manager, "There is always news about safety, and some of that news will be bad." Hence, audits are a fundamental component of any safety management system.

The SEMS rule

SEMS Rule	Discussion
11.1 General	
SEMS is a management system, and so should be audited on a regular basis. The scope, frequency, management and reporting of the audits should all be considered. The BOEMRE/SEMP requirements are typical of any audit.	Activities and areas for the audit should be defined in the SEMS Management Document. It is important to give this topic sufficient consideration. It will include physical boundaries (for example, whether subsea pipelines are included), and organizational issues such as the relationship with contractors. Much of the information will be stored at onshore facilities. However, the safety-critical information that on-board personnel may need – often in a hurry – should be identified. Information of this type should probably be available in hard copy format, because, during an emergency, it is more than likely that normal electronic data management system will not be functioning properly. Audits should always be conducted by a team The team will evaluate a mix of facility documents, interviews with key personnel and field observations. Audit reporting covers not just the structure and content of the report itself, but also a list of who is to receive a copy of the report, and how action items are to be followed up on.
11.2 Scope	
The scope of the audit includes a review of the management elements and checking the record-keeping. The audit should also try to identify areas of improvement.	Further discussion as to how the audit can help identify areas of improvement is provided in Chapter 5.
11.3 Audit Coverage	
The standard recognizes that auditing can be conducted on a sample of facilities and operations. However, that sample should cover a cross-section of personnel, equipment and regional areas. The standard calls for 'at least fifteen percent (15%) of the facilities' to be audited.	The standard calls for the audit to cover a broad range of information sources. It also requires that the audit cover different types of facilities and technologies. The requirement for 15% coverage is substantial. However, this number may need to be increased if the 'sufficient deficiencies' are identified.

SEMS Rule	Discussion
11.4 Audit Plan	
The standard requires that each audit should be conducted according to a written plan. The plan should cover the items and issues covered in the previous sections.	Like any other management activity, an audit should be properly planned. In particular as many as possible of the administrative details and information acquisition steps should be carried out before the audit team actually starts work. The plan should clearly define the objectives of the audit, the geographical and organizational scope of work, and the anticipated amount of time that will be needed. Audits are generally organized around the following steps: • determine the goals of the audit; • establish the audit standards to be followed; • define the scope and budget; • conduct the audit; • issue a report; • have a close out meeting; • follow up; and • provide guidance as requested.
11.5 Audit Frequency	
The standard calls for an initial audit to be carried out within 2 years, and for subsequent audits to be conducted every 4 years or less.	This section of SEMP is probably not relevant to SEMS. BOEMRE has the authority to conduct an audit at any time after November 15, 2011.
11.6 Audit Team	
RP 75 states that audits 'may be performed by personnel from within the organization and/or by external persons selected by the organization'.	With regard to having internal personnel conduct an audit this may be one key area of difference between SEMP and SEMS. BOEMRE requires the use of Independent Third Party Auditors (I3Ps). This may preclude operators from using their own personnel as auditors. Clarification on this issue is likely to be provided with the release of SEMS II.
11.7 Audit Report	
The audit report should discuss the items listed in the audit plan.	One of the challenges of process safety work is that the analyses and studies often result in a written report. But getting the report finished on time and of sufficiently high quality often turns out to be quite a challenge. The bulleted items provided in RP 75 provides a basis for a Table of Contents of the audit report. The audit report must show that there is a system for responding to findings. The system can part of the facility's overall process for tracking and responding to identified hazards.

SEMS Rule	Discussion
	Once the audit has been completed a report must be issued. An audit report should stress the reasons for conducting the audit, and should make explicit the standards against which the audit is being conducted.

BOEMRE requirements

BOEMRE has four sections to do with the topic of auditing and the follow-up to audits.

§250.1920 What are the auditing requirements for my SEMS program?

BOEMRE Requirement	Discussion
(a) You must have your SEMS program audited by either an independent third-party or your designated and qualified personnel according to the requirements of this subpart and API RP 75, Section 12 (incorporated by reference as specified in §250.198) within 2 years of the initial implementation of the SEMS program and at least once every 3 years thereafter. The audit must be a comprehensive audit of all 13 elements of your SEMS program to evaluate compliance with the requirements of this subpart and API RP 75 to identify areas in which safety and environmental performance needs to be improved.	The BOEMRE requirements are in alignment with those listed above, with the following differences. • The audit frequency has changed from once in 4 years to once in 3 years. • The audit includes 13 elements. The "General" requirements of SEMP must be covered.
(b) Your audit plan and procedures must meet or exceed all of the recommendations included in API RP 75 Section 12 (incorporated by reference as specified in §250.198) and include information on how you addressed those recommendations. You must specifically address the following items: (1) Section 12.1 General (2) Section 12.2 Scope (3) Section 12.3 Audit Coverage (4) Section 12.4 Audit Plan You must submit your written audit plan to BOEMRE at least 30 days before the audit. BOEMRE reserves the right to modify the list of facilities that you propose to audit.	One of the general features of SEMS is that companies are not required to submit a program or plan. They must simply have the program in place such that they are ready for an audit (or incident investigation). There are, however, a number of exceptions to this generalization—and this 30 day requirement is one of them.
(5) Section 12.5 Audit Frequency, except your audit interval must not exceed 3 years after the 2 year time period for the first audit.	As already noted, this requirement is over and above what RP 75 calls for. Discussed on page 221.

BOEMRE Requirement	Discussion
(6) Section 12.6 Audit Team. The audit that you submit to BOEMRE must be conducted by either an independent third-party or your designated and qualified personnel. The independent third-party or your designated and qualified personnel must meet the requirements in §250.1926.	
(c) You must require your auditor (independent third-party or your designated and qualified personnel) to submit an audit report of the findings and conclusions of the audit to BOEMRE within 30 days of the audit completion date. The report must outline the results of the audit, including deficiencies identified.	This paragraph is self-explanatory. Once more, BOEMRE is adding a timetable requirement to the original SEMP standard.
(d) You must provide the BOEMRE a copy of your plan for addressing the deficiencies identified in your audit within 30 days of completion of the audit. Your plan must address the following: (1) A proposed schedule to correct the deficiencies identified in the audit. BOEMRE will notify you within 14 days of receipt of your plan if your proposed schedule is not acceptable. (2) The person responsible for correcting each identified deficiency, including their job title.	If the facility has an overall hazards tracking system, as discussed on page 157, then its workings should be included in the audit report. (1) The agency does not say what happens if they themselves fail to meet the 14-day requirement. (2) Identification of responsible parties is an integral part of the Hazards Register (page 157).
(e) BOEMRE may verify that you undertook the corrective actions and that these actions effectively address the audit findings.	How this requirement is to be met is not specified.

§250.1924 How will BOEMRE determine if my SEMS program is effective?

BOEMRE Requirement	Discussion
(a) BOEMRE or its authorized representative may evaluate or visit your facility to determine whether your SEMS program is in place, addresses all required elements, and is effective in protecting the safety and health of workers, the environment, and preventing incidents. BOEMRE or its authorized representative may evaluate your SEMS program, including documentation of	This paragraph is self-explanatory: the agency has the legal right to evaluate the effectiveness and application of a SEMS program at any time of its own choosing.

BOEMRE Requirement	Discussion
contractors, independent third-parties, your designated and qualified personnel, and audit reports, to assess your SEMS program. These evaluations or visits may be random or based upon the OCS lease operator's or contractor's performance.	
(b) For the evaluations, you must make the following available to BOEMRE upon request: (1) Your SEMS program; (2) The qualifications of your independent third-party or your designated and qualified personnel; (3) The SEMS audits conducted of your program; (4) Documents or information relevant to whether you have addressed and corrected the deficiencies of your audit; and (5) Other relevant documents or information.	Once more, this paragraph is both self-explanatory and sweeping in its scope. When reviewing documents an auditor is likely to want to know four pieces of information: (1) The document type (2) Examples of the document (3) Where the documents are retained (4) The retention period
(c) During the site visit BOEMRE may verify that: (1) Personnel are following your SEMS program; (2) You can explain and demonstrate the procedures and policies included in your SEMS program; and (3) You can produce evidence to support the implementation of your SEMS program.	Although the SEMP/SEMS standard does not include an *Employee Participation* element in the way that OSHA's PSM standard does, this paragraph has something of the same effect.
(d) Representatives from BOEMRE may observe or participate in your SEMS audit. You must notify the BOEMRE at least 30 days prior to conducting your audit as required in §250.1920, so that BOEMRE may make arrangements to observe or participate in the audit.	At the time of writing, the practicalities of these requirements are not clear, because it would appear as if BOEMRE does not actually have enough personnel to meet the considerable requirements of this paragraph.

§250.1925 May BOEMRE direct me to conduct additional audits?

BOEMRE Requirement	Discussion
(a) If BOEMRE identifies safety or noncompliance concerns based on the results of our inspections and evaluations, or as a result of an event, BOEMRE may direct you to have an independent third-party audit of your SEMS program, in addition to the regular audit required by §250.1920, or BOEMRE may conduct an audit.	This paragraph contains one of the many BOEMRE references to "independent third-party audits."

BOEMRE Requirement	Discussion
(1) If BOEMRE direct you to have an independent third-party audit, (i) You are responsible for all of the costs associated with the audit, and (ii) The independent third-party audit must meet the requirements of §250.1920 of this part and you must ensure that the independent third-party submits the findings and conclusions of a BOEMRE-directed audit according to the requirements in §250.1920 to BOEMRE within 30 days after the audit is completed. (2) If BOEMRE conducts the audit, BOEMRE will provide a report of the findings and conclusions within 30 days of the audit.	The 30-day requirements in both of these paragraphs are likely to be a challenge. Writing the audit report, having it checked, and then issued can take quite a lot of time.
(b) Findings from these audits may result in enforcement actions as identified in §250.1927.	
(c) You must provide the BOEMRE a copy of your plan for addressing the deficiencies identified in the BOEMRE-directed audit within 30 days of completion of the audit as required in §250.1920.	

§250.1926 What qualifications must an independent third-party or my designated and qualified personnel meet?

BOEMRE Requirement	Discussion
(a) You must either choose an independent third-party or your designated and qualified personnel to audit your SEMS program. You must take into account the following qualifications when selecting the third-party or your designated and qualified personnel: (1) Previous education and experience with SEMS, or similar management-related programs. (2) Technical capabilities of the individual or organization for the specific project. (3) Ability to perform the independent third-party functions for the specific project considering current commitments. (4) Previous experience with BOEMRE regulatory requirements and procedures. (5) Previous education and experience to comprehend and evaluate how the company's offshore activities, raw materials, production methods and equipment, products, byproducts, and business management systems may impact health and safety performance in the workplace.	These requirements are self-explanatory.

BOEMRE Requirement	Discussion
(b) You must have procedures to avoid conflicts of interest related to the development of your SEMS program and the independent third-party auditor and your designated and qualified personnel. (c) BOEMRE may evaluate the qualifications of the independent third-parties or your designated and qualified personnel. This may include an audit of documents and procedures or interviews. BOEMRE may disallow audits by a specific independent third-party or your designated and qualified personnel if they do not meet the criteria of this section.	

Audit protocol

Audit protocols are generally structured on the following lines:

(1) The regulatory requirement of statement is inverted and turned into a question.

(2) The auditor aims to find out if the requirement is being met. He or she is looking to create a simple "Yes/No" answer, with a partial "Yes" being the same as "No."

(3) The auditor usually obtains information from one of three sources:

(a) A written document.

(b) Discussions with an individual at the facility.

(c) Observation of field activities.

(4) Ideally, at least two of the above information sources should be used for each question, and the answers should be cross-checked with one another.

(5) There will be a space for the auditor to enter notes and comments.

(6) There will be links to reference material and other supporting information (in the case of SEMS this will typically include the SEMP audit guidance).

In the case of a SEMS audit, the above model has to be modified slightly. Many of the questions are based on the SEMP standard, which is more discursive than a typical regulation. Therefore, an audit question has to be developed from the general statement.

An example of an audit protocol for one of the management of change requirements is shown in Figure 4.9 (it has been made physically smaller than an actual form in order to fit on to a book page).

The following comments apply to Figure 4.9.

Single question

The audit question is concerned with just one topic: new production or process facilities. This particular section of the MOC standard actually contains a total of 13 subparagraphs. Therefore, 13 audit protocols will have to be created.

4.2 Change in Facilities a. Is Management of Change used during construction of new production or process facilities?		Y	☐
		N	☐
		N/A	☐
Facility			
Area			
Documents Reviewed	Document Title		Date
Persons Interviewed	Name	Title	Date
Field Observations			
Notes			
Reference Material			

FIGURE 4.9

Example audit protocol.

Answers

Three answers are possible: "Yes," "No," and "Not Applicable." There should be sufficient information on the audit form and its attachments to justify the selection of the answer. If "N/A" is selected, the audit form should explain why.

Frequency of audits

The SEMS rule requires that facilities conduct their first audit within 2 years and at least once every 3 years thereafter. In practice, those companies that do not have a SEMS program that is in good condition will probably start with an audit which will serve as a gap analysis.

Audit team

The SEMS rule makes it clear that the qualified, independent third-party auditors can be used. Details regarding the audit program were provided earlier in this chapter. BOEMRE says that they "would be more likely to participate as an observer in the case where the third-party auditor is the same as the contractor who developed the SEMS program."

Cost of audits

The development and application of a SEMS audit program is discussed on page 252. Naturally, the scope of an audit will depend on a wide range of factors such as size of the facility, its age and the maturity of the SEMP/SEMS program. However, it is unlikely that a full audit (including preparation, execution, and reporting) would take less than 6 man-weeks, or 240 hours. If the rate of $100 per hour used above is used (an optimistic assumption) then the cost of an audit would be $24,000—once more excluding the cost of any follow-on work.

RECORDS AND DOCUMENTATION

The SEMS rule requires that companies keep complete, accurate, and up-to-date records of their work. This element is one area in which the offshore SEMS standard differs from OSHA's PSM standard for onshore chemical and refining facilities.

The SEMS rule

SEMS Standard	Discussion
12.1 General	
This section discusses the management of records and documentation. It states that, 'Records or documentation may be in either paper or electronic form'.	The provision that, 'the documentation does not have to be retained in a separate file...' reduces the amount of effort needed to meet this requirement. However, it also means that a comprehensive indexing system will be required. It is a good idea to keep critical documents that may be needed urgently (generally to do with emergency response) on paper as well as electronically. Then, if there is

SEMS Standard	Discussion
	a total power failure, that information still remains accessible.
12.2 Documentation/12.3 Records/12.4 Related Documentation	
Documentation is an integral part of the whole SEMS program. The documentation needs to describe not just the 12 technical elements of SEMS, but also the manner in which they interact with one another.	Management of the interaction between the elements of SEMS is both important and difficult. The interaction processes need to be properly documented and understood.
12.5 Record and Document Control	
This section discusses the manner in which records and documents can be located and retrieved.	One of the greatest challenges regarding records and documentation offshore is that most of the information will be stored in offices onshore. In principle, any information that is stored electronically should be accessible to personnel working offshore; nevertheless, there are likely to be problems with finding data in a timely manner. Information can be stored in any form: electronic, paper and intranet are all acceptable (Levine, 2011). The facility or company will need a policy as to how long documents are to be retained, and how to dispose obsolete documents.

REFERENCES

CSB (U.S. Chemical Safety and Hazard Investigation Board): Regulatory Approaches to Offshore Oil and Gas Safety, *Public Hearing*, December 2010.

Glenn David G: Job Safety Analysis. Its Role Today. *Professional safety.* March 2011.

NSC (2009). *Accident Prevention Manual for Industrial Operations.* Fifth ed. 2009.

Parker W. *SEMS Workshop Opening Statement*, Offshore Operators Committee, 2009 09 02.

United States Steel Corporation: *Principles of Accident Prevention in United States Steel*, third ed., Pittsburgh, Pennsylvania, 1968.

CHAPTER

Implementing SEMS

5

CHAPTER CONTENTS

INTRODUCTION

The Bureau of Ocean Energy Management, Regulation and Enforcement (BOEMRE) does not provide detailed guidance as to how a Safety and Environmental Management System (SEMS) program is to be implemented because the standard is fundamentally nonprescriptive. Each company and facility will need to develop a SEMS compliance plan that addresses their own needs and circumstances. Some thoughts as to how this can be done are presented in this chapter.

TYPES OF OPERATORS

BOEMRE divides oil and gas operators into three broad categories depending on their level of activity: high, moderate, and low. The division depends on the thousands barrels of oil equivalent (MMBOE) produced per day. BOEMRE reviewed data that had been submitted by many of the operators, and published the summarized information shown in Table 5.1.

Table 5.1 Types of Operators

Size	Production Rate	Number of Operators	Percentage with SEMP
High Activity	≥ 10 MMBOE	13	100
Moderate Activity	1 MMBOE < 10 MMBOE	41	71
Low Activity	<1 MMBOE	76	37

Ian Sutton: Offshore Safety Management. DOI: 10.1016/B978-1-4377-3524-6.00005-8
Copyright © 2012 Ian Sutton. Published by Elsevier Inc. All rights reserved.

The first conclusion to be drawn from Table 5.1 is that high activity (usually the larger) companies should have little trouble meeting the SEMS rule because they are all in 100% compliance with the Safety and Environmental Management Program (SEMP). This does not mean, however, that these companies do not have some make-up work to do. As discussed in the previous chapter, when it created SEMS, BOEMRE added a considerable amount of material to SEMP. It is possible that some of the high-activity companies may find that there are areas in which they still do not meet the requirements of SEMS.

Table 5.1 shows that there are 41 moderate activity operators. They report good compliance with SEMP. But they are also likely to have a considerable amount of work to do to comply with the SEMS standard due to BOEMRE's additional requirements.

The largest number of operators is in the low activity category—most of these operators do not have an effective SEMP. Given the very short amount of time available to implement a SEMS program, these companies will be challenged to meet the deadline.

The information shown in Table 5.1 is self-reported—it was not verified through an independent audit. Therefore, it is possible that some companies, both large and small, may have indulged in a little unjustified optimism or wishful thinking regarding their current status, and that they are not actually in full compliance with the SEMP standard as they report. It has also been reported (Bromwich, 2011) that the percentage of companies meeting the SEMP requirements has been slipping.

The following general statements can be made about companies of different sizes and the SEMS work that they are likely to have to do in the first few years of the implementation of SEMS.

High-activity companies

Even though these companies already have complete and effective Safety Management Systems (SMS), usually in the form of a SEMP, they will need to conduct a gap analysis to show that they have met the letter as well as the spirit of the law, particularly if their SMS uses a different structure from that of the SEMP (companies that use safety cases would be an example of such a difference). In all these cases it will be necessary to create a map that shows how the elements of their program match SEMS requirements. Figure 5.1 is an illustration of such a map.

The left side of Figure 5.1 shows how the Operator maps to SEMS; the right is the same for the Contractor. This approach avoids having to create multiple maps for each Operator and Contractor.

For example, paragraph §250.1914 (e)(1), which is shown on the left side of the SEMS rule, states,

> *(e) You must:*
>
> *(1) Perform periodic evaluations of the performance of contract employees that verifies they are fulfilling their obligations, and*

The above Statement is inserted into the left side of Figure 5.1. The arrows point to the Sections of the Company's own program that match the details of that requirements.

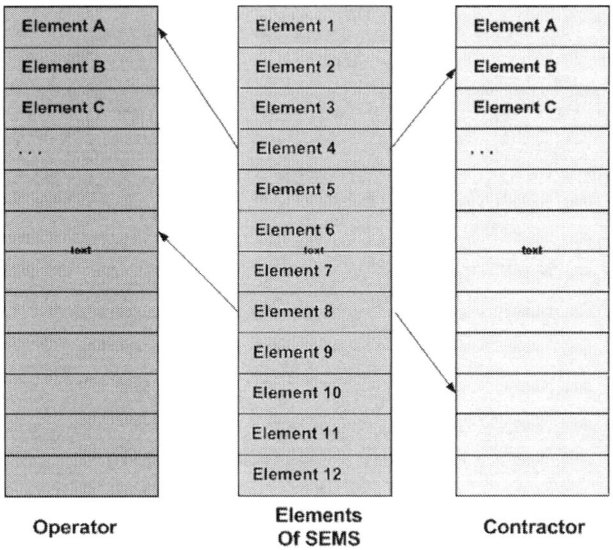

FIGURE 5.1

SEMS Map.

Moderate activity companies

Typically, the middle range of companies will have a good SEMP program, but there could be serious gaps. Many of these companies will have a lot of catch-up work to do. Therefore, in addition to creating a mapping process such as that shown in Figure 5.1 they will probably need to prioritize their work using a risk-ranking methodology.

Low activity companies

Based on the results shown in Table 5.1, many of the smaller operators have a lot of work to do before they have a full and effective SEMS program. Realistically, they are not going to have a complete program by the 2011 deadline. It is shown below how other regulations and standards similar to SEMS took about five years to implement in those companies that had no existing program. Some of the low activity companies may find that they need five years to implement SEMS. For companies such as these it is suggested that a risk-based program be used.

Contractors

The responsibility for implementing and managing SEMS lies with the operating companies. Contracting companies—even very large ones—manage their safety programs under the direction of the operator. In practice, most contractor companies do have their own programs, so it is useful to analyze them in the same manner as was done above for the operating companies.

Many of the smaller contractors will only need a partial SEMS-type program. For example, a company that provides just painting services may only need to address the elements of SEMP that are highlighted in Table 5.2.

Table 5.2 has three columns. The first column lists the elements of SEMS (excluding "General"). The second column shows whether or not this painting contractor believes that his company has to have a safety program to do with that element. If he does, then

Table 5.2 Example of SEMS Strategy for Small Contractor

SEMS Element	Applies	Notes
1. Safety and Environmental Information	Y	MSDS information to do with paint, solvents and other chemicals should be provided to the Operator. Note: MSDS information is not explicitly identified in RP 75. However, RP 14C requires MSDS for hazards analyses.
2. Hazards Analysis	S	The contractor may participate in those parts of a HAZOP to do with fire retardant equipment.
3. Management of Change	S	
4. Operating Procedures	N	
5. Safe Work Practices	Y	Safe work practices for the handling and use of paint and solvents should be incorporated into the Operator safe work program. All employees should be trained in the Operator's Safe Work Practices.
6. Training	N	
7. Assurance of Quality and Mechanical Integrity of Equipment	Y	Equipment used for painting should meet all mechanical integrity standards.
8. Pre-Startup Review	N	
9. Emergency Response and Control	Y	Special emergency response procedures for paint and solvents should be supplied to the operator. All employees should be trained in the Emergency Response and Control procedures for the facilities on which they are working.
10. Investigation of Incidents	S	
11. Audit of Safety and Environmental Management Program Elements	N	
12. Records and Documentation	S	Safety information, such as MSDS, should be supplied to the operator to be included in the overall records and documentation.

that safety program will have to be bridged to the operator's SEMS. The letters shown have the following meaning:

- "Y"—the contractor has to have a safety program that is to be bridged to the SEMS.
- "S"—the contractor will support the operator as required. For example, if the painting tasks are considered to be a system change, the contractor will participate in, but not lead, a Management of Change (MOC) analysis. Similarly, for hazards analysis, the painting contractor would participate in a Hazard and Operability (HAZOP) as requested, but would not be expected to lead or initiate that activity.
- "N"—the contractor neither needs a safety program for this element, nor is he likely to have to participate in operator activities.

TIMING

BOEMRE gave operators just one year to have a full SEMS program up and running—with the deadline being November 15, 2011. Table 5.1 shows that over half the operators did not have a full SEMP or SEMS in the year preceding the deadline, and, realistically, most of these companies will head into the year 2012 with those programs only partially complete. Even the larger companies (the high activity operators in Table 5.1) are likely to find areas that need attention. Regardless of the program that is adopted, three activities should be carried out right away:

- The P&IDs must be brought up to date. These drawings are foundational to almost all of the elements of SEMS, and so they must receive priority.
- Changes are always being made to systems and equipment. Many of these changes will require that the P&IDs be updated. Therefore it is critical that the Management of Change (MOC) system be up and running at the start of the SEMS project.
- BOEMRE has made it clear that companies must have their hazards analyses completed by November 15, 2011.

The moderate and low activity companies that do not have an existing SEMP program, or whose program is not close to being complete, are in a less-than-enviable position. One year is simply not enough time to develop and implement a full SEMS program. The amount of time that is needed will depend on many factors, such as the current status of the program and the complexity of the operations. However, assuming that a company is starting with nothing, a working number would be five years.

One historical precedent that justifies the above estimate comes from the implementation of the Occupational Safety and Health Administration (OSHA) Process Safety Management (PSM) standard. As discussed in the previous chapter, there are strong similarities between PSM and SEMS. Therefore, the time that companies needed to implement PSM probably provides sensible guidance for the same activity with SEMS. In 1992, OSHA told companies that they had to have most of the elements in place immediately. However, the agency did allow for a 5-year implementation period for certain key activities such as the completion of Piping and Instrument Diagrams (P&IDs) and hazards analyses. Consequently, PSM became—unofficially—a 5-year standard. And, in general, that was about the right amount of time that a committed company needed to be in compliance.

Another relevant benchmark comes from a public hearing that the Chemical Safety Board held on September 15, 2010. Authorities on the implementation of safety cases in both the United Kingdom and Norway said that it took about five years to develop a representative program.

DESIGNING A SEMS PROGRAM

As has been repeatedly stressed, there cannot be a one-size-fits-all SEMS program that applies uniformly across the board. Each company and each facility is unique and faces its own distinct challenges. Nevertheless, it is possible to outline a generic SEMS Implementation Program that can be used as the basis of facility-specific programs. Indeed, a SEMS program can be organized like any other management program in the following manner:

Step 1—Determine the objectives of the program.

Step 2—Create the organization.

Step 3—Develop a plan.

Step 4—Implement the plan.

Step 5—Audit and improve.

Step 1—Determine the objectives

With any management program the first task is to determine the objectives. With regard to SEMS the first and most obvious goal is to create a program that meets the requirements of the law.

However, it would be disappointing were management to stop there. The development of a SEMS program provides a company or facility with an opportunity to review its overall safety and environmental goals, and determine if a new and higher level of performance can be achieved. Another goal of the SEMS program for a large company could be to standardize safety management practices across all the company's activities and facilities.

All objectives must be measurable, and they should be understood and accepted by everyone working at the facility or on a project. They should be as concrete and specific as possible and the level of detail associated with them should be tightly defined.

If the goals of the program are not properly defined, then scope creep may occur. As any project progresses it is tempting to modify the goals to reflect new ideas and to address problems that had not been considered, thus increasing the scope of work. This is a particularly serious problem with regard to SMS because there is always the feeling that any new safety problem has to be addressed and incorporated into the program—not to do so would appear to be irresponsible. Yet every additional activity will lengthen the overall time needed for completion, and will increase costs.

Step 2—Create the organization

The next step in the development of a risk management program is to set up an organization. This is often done through use of a steering committee, subcommittees, and an SEMS coordinator, as shown in Figure 5.2.

Corporate

At the top of Figure 5.2 is the company's corporate governance. The executives who work at this level set the company's overall goals and strategy. They are the ultimate customers of the SEMS program, and it is they who fund the SEMS activities.

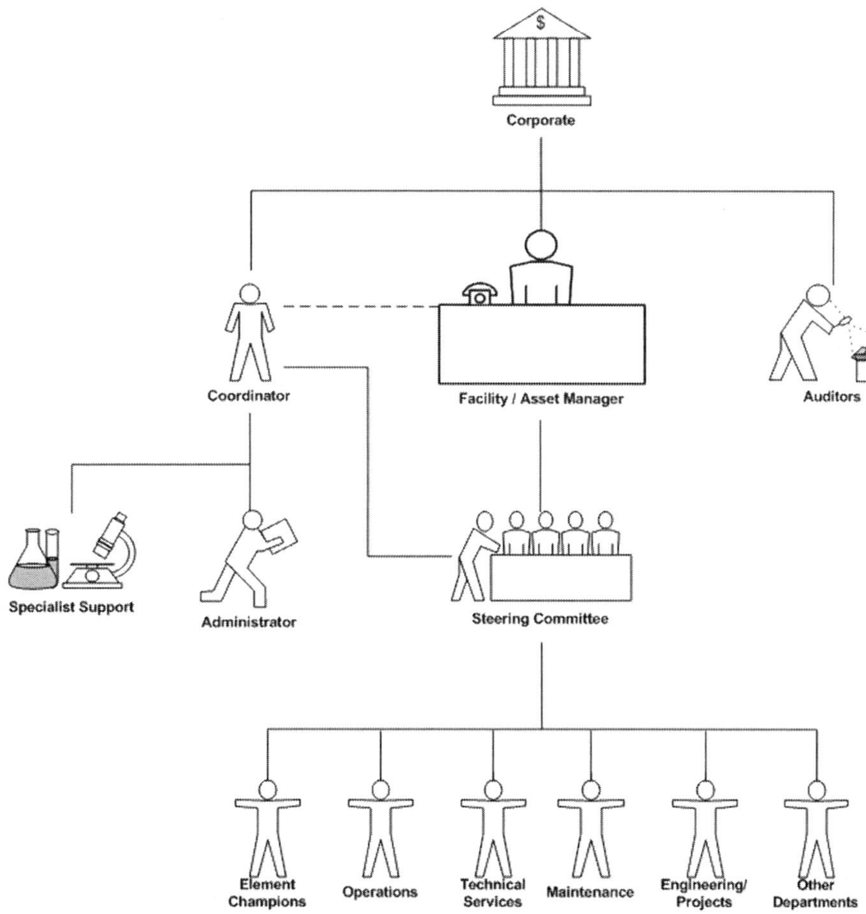

FIGURE 5.2

Representative Organization.

If a company has more than one facility then they may be able to achieve economies of scale by developing programs for training or operating procedures at a corporate level. These can be taken by the individual facilities and modified to reflect their particular needs and circumstances.

Facility management

Reporting upwards to the corporation is the manager who is in charge of the drilling rig or production platform for which the SEMS program is being implemented. He or she is in direct line of command and his or her orders must be followed.

The facility manager is not likely to spend more than a few hours a week on the SEMS program. However, he or she will need to have frequent communication with the SEMS team to make sure that all plans are being followed. The manager must also direct the interfaces with other facility activities. For example, there is obviously a good deal of overlap between those in charge of the Mechanical Integrity element and the managers and supervisors in the maintenance department.

The manager will be responsible for the funds spent on SEMS, and will have to report to his or her corporate managers as to overall progress. He or she will also be closely involved with audits and hazards analyses, and with the implementation of findings and recommendations.

Steering committee

Many companies elect to have their safety management programs organized and run by a steering committee. This committee is charged with developing detailed goals and targets and for actually implementing SEMS. The committee should be chaired by the facility manager. Typically, the steering committee will be composed of the following persons:

- Facility manager or Offshore Installation Manager (OIM)
- Operations manager on a production platform
- Toolpusher or equivalent on a drilling rig
- Maintenance manager
- Engineering/Technical manager
- Safety supervisor or HSE advisor
- SEMS coordinator

Subcommittees

Reporting to the steering committee are subcommittees, one for each of the major elements of the program. Each of these subcommittees will be responsible for the development of a detailed policy for their particular element of the standard within the overall company guidelines, for ensuring that it is properly implemented and maintained.

The chair of each of these subcommittees should, where feasible, remain the same person for as long as possible in order to provide continuity. However, there are some advantages to rotating the chairmanship of these subcommittees in order to give different people a chance to work on the different elements of the risk management program.

Some of the subcommittees will have a lot of work to do, and will be active all the time. Mechanical Integrity certainly falls into this category. Other subcommittees, such as Pre-startup Review, may find that their workload and level of activity will vary depending on current operating conditions; they will be busy at turnarounds but not at other times.

Regardless of the structure of the subcommittees, the responsibility for the management and control of each of the elements will always lie with the pertinent departmental managers. For example, operating procedures will be written and updated by the operations department, engineering information will be controlled by the engineering manager, and mechanical integrity will be under the direction of the maintenance manager. These managers decide on how the program is to be implemented in detail, how training is to be conducted, and how progress is to be measured. They will usually chair the pertinent subcommittees, but that is not a requirement.

An example of some of the detailed subcommittee requirements is provided in Table 5.3 for the topic of training.

SEMS coordinator

For all but the smallest operators, it will usually be necessary to appoint a SEMS coordinator. This person is responsible for the day-to-day management of SEMS work. He or she will report to the facility manager and will participate in as many of the

Table 5.3 Organizational Responsibility for Training

	Method of Implementation	Location of Records	Responsibility
Initial Training Required before operating a process except for "grandfathering" those already involved in operating a process	Mix of classroom and field observation.	Training department. (Documentation requires employee ID, date, and means used to verify understanding.)	Training department.
Refresher Training Refresher training required at least every 3 years, but the actual frequency is determined by the employee and his/her manager.	Classroom and field training.	Training department.	Operations and maintenance departments.
Emergency Response and Troubleshooting Training All operations personnel should have at least 4 hours a year of training in emergency response.	Emulators and simulators.	Training department.	Operations department.

subcommittees as time permits. The responsibilities for this person include the following:

- Keep the SEMS program and plan up to date, and adjust these documents as working conditions change.
- Report to management on progress, and areas of concern.
- Report on costs.
- Work with outside parties such as auditors from BOEMRE.
- Find the right people to lead and conduct the SEMS work. The biggest challenge here is that the people who know the most and who can make the biggest contribution are typically very busy elsewhere, and in demand by many other people.
- Train participants in the elements of SEMS. For example, most persons at the facility at supervisor level or above will need to be very familiar with the company's MOC program.

The coordinator is usually a very busy person (and often has other responsibilities). Therefore, he or she should be supported by an administrator who carries out tasks such as managing the risk register and scheduling training.

Consultants

Many companies will choose to hire consultants to help them set up the SEMS program. These consultants may also serve as the independent third party auditors (I3Ps), although BOEMRE has stated that they are more likely to double check a program if the contractor who helped create the program is also the program's auditor.

It is useful to make a distinction between consultants and contract workers. A consultant is someone who works with management to help create and guide the SEMS program. Often he or she will possess expert knowledge in specific areas such as regulatory compliance. They provide the specialist support shown in Figure 5.2.

A contractor, on the other hand, is hired to execute a specific task such as leading a hazards analysis or writing a set of operating procedures. A contractor is not asked to help with policy setting.

RACI chart

A Responsible, Accountable, Consult, and Inform (RACI) chart, an example of which is provided in Figure 5.3, outlines the roles of different persons on different parts of a project.

Signature authority

An important part of setting up the organization consists of deciding who has signature authority, and what their signature signifies. This is not a trivial consideration. Documents are often held up because they are stacked up in someone's email inbox or in a squad check room. Therefore, signatures should only be called for when they are truly adding value.

Another difficulty to do with the signature process is that too many people are being asked to sign the document. Therefore, it takes a long time for that document to wend its way through the system. Second, the people who know the unit well, and whose

	Activity 1	Activity 2	Activity 3	...
Person A	A	A	C	
Person B	R	R	A	
Person C	C	—	—	
Person D	I	I	C	

The letters in Figure 5.3 are explained below.

Responsible	R	Expected to actively participate in the activity and provide contributions.
Accountable	A	Person ultimately responsible for the results.
Consult	C	Persons having expertise that they can contribute.
Inform	I	People who are affected by the activity, but do not participate in it.

FIGURE 5.3

Sample RACI Chart.

signature is, therefore, of the highest importance, are also the people who are likely to have the most to do with other projects, or who are busy with day-to-day operations. It can be difficult for them to make sufficient time available to review the procedures and documents with the care and thoroughness that is required.

In practice, there are usually just three levels of signature authority that really matter.

The first signature is from the person(s) who actually prepared the document that is being reviewed. Their signature states that the document is accurate, complete, and useable, within the extent of their capabilities.

The second signature is that of a reviewer who knows the process being described extremely well indeed. This person will often be a supervisor or senior technician. Their signature states that, to the best of their knowledge, the written procedure or document is accurate.

The third signature is that of a manager. It is unlikely that he or she will know the process well enough to comment on the technical content. What his or her signature states is that management systems for preparing and writing documents are in place, that the document being signed addresses all appropriate regulations and standards, and that these systems were followed during the execution of this assignment.

The SEMS manual

Experience with PSM systems has shown that the use of a program manual—in this case a SEMS manual—can help management keep control of the program. This document, which would be part of the recordkeeping requirement of SEMS, would show where all critical information is located. It would also serve as a structure for carrying out audits.

The manual can be divided into two sections: General and Management Elements. The General section describes the overall SEMS policy; the Management Elements section describes how each of the specific elements are being implemented and audited.

General section

Items in the General section could include the following:

- Safety and environmental policy
- Planning
- Implementation and operation
- Verification and corrective action
- Management review
- Continual improvement

Management elements

A section in the manual will be created for each of the elements of SEMP, as listed in Table 1.3. Some of the topics that can be discussed for each of the elements are shown below:

- Introduction
- Objectives
- Regulations (SEMS and others)
- Industry Standards
- Company Standards
- Audit Protocols
- Employee Participation
- Administration
- Equipment Items Covered
- Personnel
- Use of Outside Companies
- Project Management
- Phases of the Program
- Budget
- Schedule

Step 3—Develop a plan

Having determined the overall program objectives and having created an organization, the next step is to develop a detailed SEMS plan. Key elements in the plan are as follows:

- Management style
- Metrics and baseline

- Preliminary activities
- Plan structure
- Budget
- Schedule

Management style

The development of a SMS such as SEMS can be managed either top-down or bottom-up (or usually, a combination of the two). In general, it is best to use a top-down approach because that way management has an overall grasp as to what is going on and how much progress is being made. Using such an approach a high-level management structure is developed, and detail is added on a progressive basis. (A detailed discussion to do with the development and management of a full SMS is outside the scope of this book. However, this issue is discussed in much greater detail in Chapter 16 of *Process Risk and Reliability Management.*)

A practical difficulty with this approach is that, in the early phases of the project, a substantial amount of time and money is being spent, but there is little progress to show for it. Also, it is not always clear if the final product will meet the facility's requirements. For this reason, it often makes sense to develop some pilot projects at an early stage. A few final products for each element are prepared, and then circulated for comment. This gives everyone a chance to make suggestions.

Some elements of the standard, such as operating procedures and mechanical integrity, lend themselves well to pilot projects. Other elements, such as Pre-startup Reviews, are less amenable to this approach.

Metrics and baseline

The effectiveness of any management program—including SEMS—can only be measured through the use of system metrics. It is not enough, for instance, to say that work on the operating procedures is "progressing well." The person responsible for that element should be able to report on progress with a statement such as "The operating procedures are 63% complete."

Program creation

The first metric will be that to do with the creation of the program. Table 5.4 illustrates how this can be done.

The first column in Table 5.4 lists the elements of the SEMS program.

The second column shows the man-days that it is estimated will be required for completing each element. In the case of training, the estimate is for 580 man-days, i.e. this is the amount of time budgeted for bringing all workers into compliance with the requirements of the training element of SEMS.

The third column shows the number of man-days of effort that have actually been expended up to this point in time. For training, this value is 340.

The fourth column shows the percentage completion for each element, with training therefore being at 59%.

The fifth column is the normalized fraction for that element, i.e. the percentage completion weighted for the contribution that each element makes. Therefore,

Table 5.4 Metrics (man-days)

	Required	Used	Fraction Complete	Normalized Fraction
Safety and environmental information	20	11	55%	0.3%
Hazards analysis	460	96	21%	2.7%
Operating procedures	350	40	11%	1.1%
Training	580	340	59%	9.6%
Pre-startup review	10	10	100%	0.3%
Assurance of quality and mechanical integrity of equipment	1260	720	57%	20.3%
Safe work practices	300	280	93%	7.9%
Management of change	120	80	67%	2.3%
Investigation of incidents	20	20	100%	0.6%
Emergency response and control	130	120	92%	3.4%
Audit of safety and environmental management program elements	40	10	25%	0.3%
Records and documentation	250	99	40%	2.8%
Total	**3540**	**1826**	**31%**	

although Incident Investigation is 100% complete, it contributes only 0.6% to the overall total because it constitutes such a small fraction of the overall effort.

Ongoing program

A program similar to that shown in Table 5.4 should be developed for the SEMS program once implemented. Some of the ongoing work can be estimated with some confidence. For example, management by the SEMS coordinator can probably determine how many hours will be spent on hazards analyses each year. However, other elements are much more difficult to forecast. For example, the level of effort to be spent on the Investigation of Incidents will inevitably depend on the number and severity of those incidents. The facility's incident track record may give some guidance, but the reality is likely to be quite different.

Plan structure

Because SEMP and SEMS are, at root, performance-based standards, a sensible way of addressing their requirements is through the development of a risk management program. Moreover, a risk-based program is a virtual necessity for the large number of companies who are going to take months, even years, to come into compliance.

An example of a generic implementation program is shown in Figure 5.4 (Sutton, 2010b).

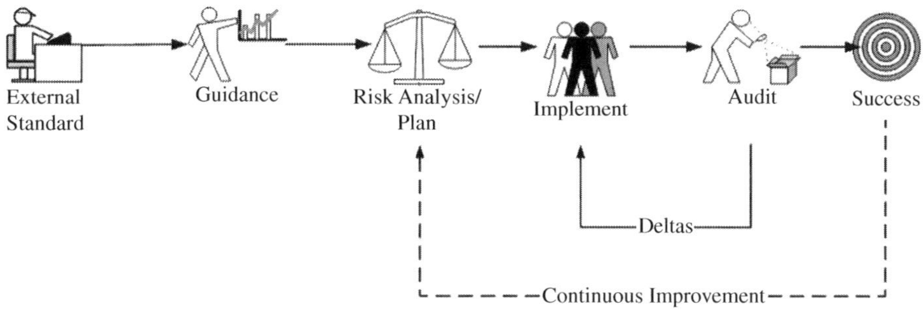

FIGURE 5.4

Implementation Program.

The process starts with the introduction of an external standard such as the SEMS rule. Generally, standards and regulations do not provide enough detail for a facility to develop a compliance plan—largely because operating conditions vary from location to location. It is not possible to have a one-size-fits-all rule. Therefore, facility management needs to develop more detailed guidance as to how the SEMS requirements at their location are to be addressed. Such guidance can cover topics such as follows:

- The frequency of inspection of equipment items
- The level of detail of operating procedures
- The process for managing changes

The next step is to create a Management Plan that shows what actions are to be taken to ensure compliance with both the rule or standard and the corresponding guidance. If the company uses specific risk management standards, such as Consequence/Frequency matrices, then they should be incorporated into the plan. One of the most important parts of the Risk Management Plan will be an evaluation of the different areas of operation, and a determination as to what areas should be addressed first.

The plan should then be implemented. It is important to be able to demonstrate to an auditor that the steps in the plan are being properly followed. If there is any slippage or change of plan, then those issues should be recorded, and explanations provided.

Once the plan has been implemented, an auditor evaluates what has been done and generates a gap analysis (the word "delta" is sometimes used to describe the difference between plan and performance because it sounds less critical than words such as "deficiency" or "failure"). The facility management then takes corrective action based on those findings. Once audits are no longer generating significant findings ("Success" in Figure 5.4), the risk management plan should be revised and upgraded.

The structure shown in Figure 5.4 is for a generic management plan. With regard to SEMS, virtually all companies will, as already discussed, have some type of SMS in place. The SMS may be based on SEMP, company standards or a safety case. But it will be very rare for a company to have nothing at all. Therefore, the implementation program will need to start with a Baseline Assessment that compares "where we are to where we need to be." In other words a gap analysis comparing the company's current program with the SEMS rule is needed.

Based on the gap analysis, a SEMS Implementation Program such as that shown in Figure 5.5 can be implemented.

The workflow in Figure 5.5 starts with the two applicable standards: SEMS and SEMP. These correspond to the "External Standard" shown in Figure 5.4.

Also the scope of the program needs to be considered, as discussed in the next section. Not all companies will have to have a full SEMS program. Once the scope is established and measured against the current program then a baseline can be set.

Budget

SEMS is expensive. This does not mean that it is a bad investment—but it is expensive.

Direct costs, such as hiring consultants to carry out studies, can be substantial, but the real cost is in terms of the time of key personnel. Virtually all of the SEMS elements requires that the senior persons on board a platform or rig must be involved. Hazards analyses, operating procedures, training, and MOC all need input from the most experienced and knowledgeable persons—otherwise the work will be ineffective. But these people are popular folk—they are needed in every other capacity, including the direct management of day-to-day safety. The cost of using these people on SEMS is not just the direct charges to do with their salaries and benefits, but also the opportunity cost. Every minute that they spend on SEMS is a minute that could be spent on other tasks that help the company make money.

There is no easy resolution to this dilemma. The SEMS program must be implemented because it is the law and because it constitutes good practice, but the company needs these senior people for so many other tasks.

Schedule

The nonprescriptive nature of process risk management means that it is difficult to develop schedules. This is one reason why it is particularly important to divide the work into manageable subprojects, otherwise the task will seem overwhelming. Each phase of the project (and each element within each phase) should be scheduled using normal project management techniques.

Step 4—Implement the plan

The order in which the elements (Tables 1.1 and 1.2) will be implemented will depend on the circumstances at each facility. However, the two items that should be in place at the very beginning of the project are P&IDs and MOC. Accurate and up-to-date P&IDs are important because so many other aspects of risk management are based on them. They are needed for Hazard Analyses, Mechanical Integrity, Operating Procedures, and Pre-startup Safety Reviews. MOC is equally important because change is constant. Therefore, as soon as one element is complete, something else will change, possibly affecting the item that was just completed. This means that the MOC process must be up and running very early on.

Preliminary steps

Before starting work on a SEMS there are certain basic activities that have to be completed in almost all situations. These include the following:

- Read the rule
- P&IDs and technical documents
- MOC

Read the rule

Management's first task is to read and understand the regulation, and interpret it in light of the company's own operations. A process can be followed in which each paragraph of the rule is interpreted in the context of the facility and the company in question. This is a boring and tedious exercise, but it needs to be done because, although most of the provisions in SEMS apply equally, regardless of the company or location, some aspects may need special interpretation, particularly when coordinating the SEMS program with contractors and other companies.

P&IDs and technical documents

Almost all SEMS work requires the use of P&IDs. Therefore, the first task is to make sure that the P&IDs are accurate and up to date. Unless the facility has had a good ongoing MOC program, it is likely that the P&IDs will not be completely accurate.

The traditional method of updating them is to assign technical personnel the task of conducting walk-downs, in which they check that the P&IDs are correct in every respect. They will need to check for all the information that is included on a P&ID, including the following:

- Line sizes
- Materials of construction
- Valve types and sizes
- Direction of check valves
- Instrument logic, including fail positions
- Vessel and line insulation

Management of change

Conditions and technical information on process facilities are constantly changing. Production rates change, new materials of construction are used, and instrumentation logic is modified. Every day is a new day. This means that it is critical that the MOC be up and running at the start of the SEMS work, otherwise those doing the work will never be sure if they are working with accurate and up-to-date data.

HAZOPs

Once the basic tasks described in the previous section have been carried out, most companies will start conducting hazards analyses. Assuming that the facility in question is already in operation, the HAZOP technique is probably the one to choose.

Step 5—Audit/Improve

Audits are a necessary part of any management system. Referring to Peter Drucker once more, "What gets measured gets done." The audit process has been described in the previous chapter. As findings and gaps are identified, management should take the opportunity not just to close those gaps but also to improve the management systems themselves through a process of continuous improvement. New and higher standards for safety can be set, and the program modified appropriately.

RISK-BASED APPROACH—PLAN B

In Chapter 4, it was pointed out that the SEMS rule differs from other programs of this type in that it has to be implemented in its entirety immediately. There is no 5-year ramp-up period as there was for OSHA's PSM or that was needed for safety cases. For those companies that do not have a functioning SEMS program, the deadline of November 15, 2011 means that the program is not going to be completed by any time close to the November deadline.

If a company finds itself in this less-than-enviable position then a plan B is needed—some plan that demonstrates to all parties (including BOEMRE) that they are making a good faith effort to come into compliance as quickly as possible. One way of doing this is to organize the SEMS program around a risk management plan—once the basic steps, such as getting the P&IDs up to date that were described in the previous section, have been completed.

A risk-based approach uses the systems shown in Figures 5.4 and 5.5 but prioritizes the work by levels of risk, starting with a Major Hazards Analysis.

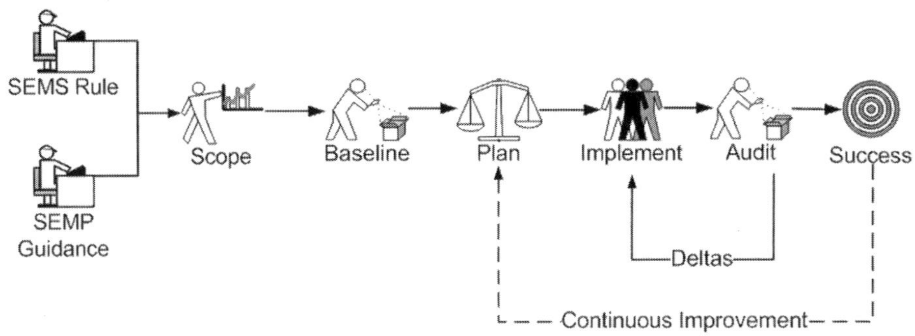

FIGURE 5.5

SEMS Implementation Program.

HIGH PRIORITY ELEMENTS

In Chapter 4, it was noted that the first version of SEMS focused on just the following four elements:

- Mechanical Integrity
- Operating Procedures

- Hazards Analysis
- MOC

The research carried out by the Minerals Management Service (MMS) suggested that these elements were the ones that contributed to the most to offshore safety incidents. An alternative risk-based approach to a SEMS program would be to focus first on these four elements, and get to the others later.

ECONOMICS OF SEMS

SEMS is a regulation—covered companies have no choice as to whether they must comply or not, regardless of cost. Nevertheless, because SEMS is such a major investment it is important to know how much it will cost, and what any benefits may be—apart from just staying in compliance with law. BOEMRE recognizes the importance of understanding the cost of SEMS—many pages in the standard discuss the topic, although as many of the commenters on the first version of the standard pointed out, it is likely that the BOEMRE estimates are very much on the low side.

The following comment and reply from BOEMRE summarizes this issue (page 63613 of the rule).

> *Most comments expressed that BOEMRE significantly underestimated the cost of developing, revising, and implementing the SEMS program. Comments also stated that BOEMRE dramatically underestimated the major new documentation and reporting burden that the rule will impose on offshore operators.*
>
> *Response: BOEMRE re-evaluated the cost burden on industry by interviewing parties experienced in the development of SEMS programs, vendors that submit information for operators, and operators with designated personnel who work on SEMS issues. Based on this information, we have increased the non-hour cost and hour burdens. Should OCS companies have documented data that shows a higher cost to industry, they may submit comments at any time on the paperwork burden as stated in § 250.199(d).*

Revenue

The true benefits of an effective SMS are to do with saving human life and protecting the environment. Of course, there is no way in which an objective, financial value can be placed on these benefits.

All accidents result in increased costs to the company involved. At the one end of the spectrum are day-to-day costs associated with issues such as workers' compensation. At the other end of the spectrum are the tens of billions of dollars that will be paid out by various companies as a result of the Deepwater Horizon catastrophe. However, it is difficult to put the potential cost savings into a normal business plan because large accidents occur only rarely.

Safety is also considered by most managers to be an effective investment. The basic idea is that if a facility operates with integrity, i.e. it operates in the way that it was meant to operate, then all facets of the operation—safety, production, environmental performance, and efficiency—will improve correspondingly. Conversely, a failure in one area will lead to a deterioration in the others. Therefore, the actions taken to improve safety will improve profitability.

Nevertheless, it is very difficult to prove such returns, so management should probably recognize that the benefits associated with an SMS are just to do with legal compliance and improving the quality of life, and leave it at that.

Costs

The costs of a SEMS program can be divided into the following four areas:

(1) Design and implementation

(2) Maintaining the program on an ongoing basis

(3) Audits

(4) Implementation of findings from activities such as hazards analyses, audits, and MOC follow-up.

BOEMRE data

BOEMRE is clearly sensitive to the costs that they are imposing on operators with SEMS. The rule contains approximately five pages of information to do with their estimates of the cost of the rule. The costs are divided according to operator size and management element (they do not provide figures for implementation costs for large operators because they all reported that their SEMS programs were already complete).

On page 63611 of the rule BOEMRE has a table which shows their estimate as to the costs of implementing and maintaining a SEMS program. It is reproduced in Table 5.5 (all values are in thousands of US dollars).

Implementation costs for high activity organizations are not reported on the grounds that all these companies have reported that they have a complete SEMP in place.

The data in Table 5.5 raises more questions than they answer. Some points of concern are as follows:

- There is no clear distinction between "partial" and "full" implementation. As has been stressed many times, nearly all offshore operators already have a SMS in place. Therefore, they can all be considered "partial," but some will be near the starting gate, whereas others may have a mature SEMP. The word "partial" is not specific enough.

- Some of the numbers pass the red-face test, other do not. The hazards analysis estimate, for example, seems defensible. Such an analysis will typically utilize a team of about six people for at least 2 weeks. An additional two man-weeks is generally needed for preparation and report writing. If a cost of $100 per person hour is assumed, then a full hazards analysis will cost $76,000—close to the estimate of $98,000 in Table 5.5.

- The figures for audits, on the other hand, seem very low. A typical audit will take say a week, and be conducted by a team of say three people. It will also take 4 hours of time of around 20 auditees (both offshore and in the onshore office).

Table 5.5 BOEMRE Estimate of Costs (US $,000)

Elements	Implementation (moderate)		Implementation (low)		Maintenance (high)	Maintenance (moderate)	Maintenance (low)
	Partial	Full	Partial	Full			
General	18	18	5	5	50	3	2
Safety and environmental information	0	22	0	8	75	12	3
Hazards analysis	0	98	0	23	300	34	14
Management of change	0	29	0	18	150	21	7
Operating procedures	0	20	-	-	-	-	-
Safe work practices	0	28	-	-	-	-	-
Training	0	30	-	-	-	-	-
Mechanical integrity	0	38	0	19	225	27	11
Pre-startup review	25	25	-	-	-	-	-
Emergency response and control	28	28	-	-	-	-	-
Investigation of incidents	20	20	10	10	95	17	3
Audits	3	3	2	2	15	6	6
Records and documentation	6	6	4	4	30	6	4
Total	**100**	**365**	**43**	**147**	**1665**	**225**	**80**

Using the figure of $100 per hour once more, the cost of a full audit comes to $20,000 or more. This far exceeds the $3000 shown in Table 5.5.

- Some of the costs are essentially indeterminate. For example, the cost of incident investigation will depend almost entirely on the number of incidents—and that cannot be predicted, except possibly by reviewing the history of incidents on the facility.
- Some of the ratios seem to be unlikely. For example, the maintenance cost for mechanical integrity is $225,000 per annum for a high activity facility, but only $27,000 per annum for a moderate activity facility. A similar disparity can be seen with regard to hazards analysis.
- The real cost of the activities depends heavily on the amount of follow-on work that they generate. For example, the $98,000 spent on hazards analyses can easily generate action items that cost upwards of $1 million to close out.
- The age of the facility is very important. For example, the technical information for older platforms may be of poor quality, partly because the drawings and other information may be on paper only, and partly because it is likely that many undocumented changes will have been made over the years.
- Some of the maintenance costs, particularly to do with hazards analysis, seem to be high.
- Some companies will charge costs to SEMS, others will not. For example, the entire mechanical integrity budget could be wrapped up with the normal maintenance costs. If, on the other hand, all maintenance costs are included in the mechanical integrity element then an ongoing cost of $225,000 to $1,665,000 is very low indeed.

The historical precedent for the investment needed for programs such as SEMS suggests that agency estimates are inadequate. For example, the cost of the OSHA's PSM program was much higher than their initial estimates.

Another way at looking at the data in Table 5.5 is to consider the overall management effort. The BOEMRE estimate for implementing SEMS on a moderate activity facility is $365,000. The program will almost certainly require a full-time coordinator with associated costs such as office space. Once the time of senior management is added in, this person could easily cost $200,000 per annum, thus leaving $165,000 to be spread over the 12 technical and management elements, i.e. giving a figure of around $14,000 each. This seems very low and does not pass the red face test.

The overall conclusion that can be drawn for Table 5.5 is that it provides a useful starting point, but does not substitute for creating a real budget. When this is done it is likely that the cost of the SEMS program will be much greater than what BOEMRE estimates.

REFERENCES

Bromwich, M. May 2011. Future of Offshore Oil and Gas Development in the U.S. Offshore Technology Conference.

Sutton, Ian. 2010a. Behavior-based Safety. www.stb07.com/occupational-safety/behavior-based-safety.html.

Sutton, Ian S. 2010b. Process Risk and Reliability Management. Elsevier, Oxford, England.

CHAPTER

Safety cases

6

CHAPTER CONTENTS

INTRODUCTION

The discussions of offshore safety up to this point in the book have focused on the management systems that are used primarily in the USA, i.e. SEMP and SEMS. However, as discussed in Chapter 1, another way of managing offshore safety is through the use of "safety cases"—an approach widely used in Europe, Australasia, and Southeast Asia.

For many safety professionals, particularly in the USA, the very phrase "safety case" is curious, but the concept behind it is quite simple. A safety case is simply the case that the designers and operators of a facility make to all interested parties that the facility is safe. It is analogous to the case that a company may have to make in its defense in a lawsuit following a major accident.

A safety case is built upon the following three principles:

(1) Those who create risks are responsible for controlling those risks.

(2) Safe operations are achieved by setting and achieving goals rather than by following prescriptive regulations.

(3) All risks must be reduced such that they are below a threshold of acceptability.

In fact, SEMS and SEMP programs share the first two goals. SEMS makes it clear that it is the operator of a facility who is responsible for its safety, and RP 75 is basically a nonprescriptive standard (although it does incorporate many prescriptive rules). One of the themes of this book is that the difference between the two approaches—SEMS and safety cases—is not as great as is sometimes thought.

Ian Sutton: Offshore Safety Management. DOI: 10.1016/B978-1-4377-3524-6.00006-X
Copyright © 2012 Ian Sutton. Published by Elsevier Inc. All rights reserved.

Safety cases are prepared for those who have an interest in the safety of the offshore facility. They can include the facility's owners, managers, the public, employees, and regulators. A safety case is not fundamentally a regulatory tool—although it is often used as such by regulators. A safety case merely aims to demonstrate to all and sundry that the facility is safe. But the main customer of a safety case is the operator of an offshore facility—all others are secondary. Once a safety *case* has been prepared and accepted by senior management, a safety case *regime* that aims to achieve the goals made in the safety case is implemented.

With regard to offshore operations, the development of safety cases in the UK sector of the North Sea received a major impetus following the Piper Alpha catastrophe. Although safety cases had been used before that event, it is probably not an exaggeration to say that the modern safety case for offshore work came into being following the publication of the Cullen report.

The principal reason for developing and implementing a safety case is, of course, to ensure that the people on a facility are safe. However, additional justifications include protection of the environment and minimizing the chance of a major economic loss (either of the facility or of the production from it).

The nature of a safety case is further explained in the following quotation (Maguire, 2006):

> *The safety case is the whole safety justification—just as is a case for the law, it comprises every appropriate piece of evidence to make a convincing argument to support some conclusion about guilt or innocence. In this case the argument concerns the safety performance of some entity or system. As a collection of evidence it needs a guide to describe how the evidence was obtained, why it was obtained and what deductions can be made from it. In a court of law, this is done by the solicitor or attorney but in a safety case this is done by the safety engineer through the safety case report. This report summarizes all the key component parts of the safety case, it makes the safety argument explicit and describes the supporting evidence. All supporting documents, analysis and results should be referenced from the safety case report. This evidence does need to be available for scrutiny, but it does not need to bulk out the safety case report.*

The Cullen report (Chapter 2) stated the following:

> *Primarily the safety case is a matter of ensuring that every company produces a formal safety assessment to assure itself that its operations are safe.*

Two aspects of the above quotation are of note.

First, the company has "to assure *itself*" that the facility is safe. At root, a safety case is developed for the facility personnel and company management—not for the outside parties. For example, operators of large and expensive deepwater facilities in the Gulf of Mexico (GoM) frequently develop analyses and reports, which are very similar to safety cases. They do this—in spite of the lack of regulatory requirements—simply to assure themselves that they have identified the factors that could lead to the loss of their very expensive facilities.

The second key feature of the above quotation is that the facility management has to develop a formal safety assessment (FSA). This means that a framework for understanding risk, and what levels of risk are acceptable, has to be developed. Just following the appropriate regulations and standards is not sufficient. This requirement means that safety cases are basically nonprescriptive and performance based. Instead of following detailed rules, the owner (duty-holder) of the facility set his or her own standards. The duty-holder's performance is then assessed against these standards. (FSAs are described in the next chapter.)

One of the justifications for the use of safety cases is that, should the worst happen, and the facility does have a serious accident, then it is likely that litigation will follow. A well-constructed and maintained safety case provides the basis of an excellent defense. Even though an accident has occurred, the safety case can demonstrate that management had given serious consideration to understanding the risk that their system posed, and that an appropriate Safety Management System (SMS) was in place.

Increasingly, companies are using the safety case methodology to go beyond just safety. They find that the management and technical systems that have traditionally been focused on safety and environmental performance can be equally effective in reducing economic losses, particularly those associated with catastrophic events. Safety cases of this type are sometimes referred to as Health, Safety, and Environmental (HSE) cases. For example, operators of very large platforms in deepwater GoM will sometimes prepare HSE cases because the costs associated with the loss of such platforms is so great. Not only does the development and application of the safety case way of thinking reduce the chance of a major event, but it can also be used to show investors, customers, insurers, and senior management that the risk associated with an expensive facility, such as a deepwater offshore platform, is at an acceptable level.

Finally, it must be stressed that a safety case is not a one-time event. A series of safety cases for each facility will be developed: one each for design, operations, and decommissioning as a minimum, in the case of production platforms. Moreover, none of these safety cases are static documents; they need to be updated whenever there is a significant change to the design or operation of an offshore facility, say from a major expansion or the introduction of a new, hazardous chemical.

Finally, a well-written safety case does not focus on just one design or a single method of operation. Instead, it examines the merits of different options and a justification that the chosen option is indeed the one that reduces risk to an acceptable level. For facilities that are already operating, the safety case should go beyond the original design information—it should incorporate actual operating experience.

SAFETY CASE DEFINITION

Based on the above discussion, a safety case can be defined as follows:

> *A documented body of evidence that provides a demonstrable and valid argument that a system is adequately safe for a given application and environment over its lifetime.*

Another definition, provided by the UK Ministry of Defence (MOD, 2004) is:

> *A structured argument, supported by a body of evidence that provides a compelling, comprehensive and valid case that a system is safety for a given application in a given operating environment.*

Yet another definition is provided by the Department of Consumer and Employment Protection of the Government of Western Australia (DOCEP, 2005).

> *A safety case regime is an objective-based regime whereby legislation sets broad safety objectives and the operator, who accepts direct responsibility for the ongoing management of safety, develops the most appropriate methods to achieve those objectives.*

It can be seen that, although the above definitions have much in common, there is not a single, agreed-upon definition as to what constitutes a safety case. Heiler (2005) states,

> *Arguably, then, the question is not what is a safety case regime—but rather what kind of safety case regime is being contemplated …*

In other words, each operator and regulator must determine the nature of the safety case for their particular situation.

DIFFERENT INDUSTRIES

First developed for the nuclear and aerospace industries, safety cases can be used in any activity or industry that poses high risk to workers or the community. One of the first examples was in the United Kingdom where the Nuclear Installations Act of 1965 required covered facilities to create and maintain a safety case in order to obtain a license to operate. Since then, other industries that have used safety cases include pipelines, railways, and mining operations. Moreover, in spite of the fact that the term "safety case" is not widely used in the USA, the same approach to the development and application of SMSs is, in fact, used in other American industries. For example, the US nuclear and space industries prepare Safety Analysis Reports (SARs) and Mission Safety Evaluations (MSEs), respectively. These documents have the same general intent and structure as a safety case. Within the onshore process industries, the safety case approach was introduced in Europe in onshore process plants as part of the "Seveso Directive" in 1986 (since replaced by the Seveso II Directive of 2003).

Although a safety case can be developed for any industrial activity, in practice, they are created only for complex systems in which the failure could have catastrophic consequences. Offshore drill rigs and platforms certainly fall into this category.

In the United Kingdom, the Seveso Directives led to the creation of the CIMAH (Control of Industrial Major Accident Hazards) regulations in 1984. These regulations required manufacturers of hazardous chemicals to create a safety report—in effect a safety case. They also had to show how the hazards were being effectively managed. (CIMAH was replaced by COMAH—Control of Major Accident Hazards—in 1999.)

FEATURES OF A SAFETY CASE

A safety case is composed of three basic elements:

(1) A facility description (FD)

(2) An FSA

(3) A description of the SMS.

These three elements must be linked to one another. For example, the FD will provide technical information that provides the basis of the FSA and the SMS. Similarly, the SMS will define what information should be included in the FD.

Because safety cases must address the unique circumstances of individual facilities, there is no single structure or format that can be uniformly applied. The risks and management activities associated with a nuclear power plant, for example, are quite different from those for a freight railroad. Therefore, the respective safety cases will be quite different from one another. Similarly with offshore facilities, no two platforms or rigs are alike; so no two safety cases will be identical.

The following are core features of a safety case:

- Duty-holder responsibility
- Participation and commitment
- Information availability
- Nonprescriptive and performance based
- Risk management system
- Management systems
- Living document
- Auditor/assessor responsibility

Duty-holder responsibility

At the heart of the safety case approach is an understanding that it is the operator of a facility—not the regulator—who decides how to ensure safe operations. The organization that prepares and manages the safety case is known as the "duty-holder." Generally, the duty-holder is the owner/operator of the facility. However, this responsibility can be delegated. For example, in the North Sea, many of the larger oil companies have sold some of their older fields to smaller organizations. These organizations, in turn, frequently do not have the internal resources develop or manage the facilities that they have bought, and to manage the corresponding safety case. Therefore, they may choose to hire a third party to act as duty-holder and therefore to manage the facility's safety case.

Participation and commitment

The active participation of all employees and contract workers is the key to the success of any safety program—including safety cases. Not only does this mean that employees are informed and trained about the safety case, but they also actively participate in its application and are encouraged to think of ways of improving system safety. The safety case becomes part of the organization's overall safety culture.

The commitment of management is also required. Given that the development and implementation of a safety case is expensive and time-consuming, management must commit the necessary funds and time of key personnel—people whose services are always in demand elsewhere in the organization—to the development and implementation of the safety case.

Information availability

The safety case contains within itself all the information that is needed to support the arguments that it presents. It is a self-contained document.

Risk management system

A safety case is not just an analytical or discussion document. It results in a risk management system that has to be implemented. Such a system generally is structured based on the following lines:

- Identify the hazards.
- Determine the level of risk associated with each hazard.
- Describe how the risks are to be controlled.
- Describe the SMS that ensures that the controls are effectively and consistently applied.
- Describe the responsibilities and activities of the system's auditors.

The risk management system usually includes a quantitative analysis, i.e. the risk associated with each of the hazards in a facility is estimated numerically and given a value—typically in the form of so many fatalities or environmental releases per 1000 years. These individual risks are then added to one another to give an estimate of the overall risk.

Discussion to do with the quantification of risk is provided in Chapter 7. If the safety case analysis does include quantification of risk, then some means of determining the acceptability of levels of risk is required, as discussed in Chapter 1.

Safety management systems

Systems for controlling risk should concentrate on management systems, not just on hardware and instrumentation. Therefore, the safety case must show that the correct management system for controlling safety is in place. Such a system is often referred to as an SMS. It is the system by which hazards are identified and risks are continually and systematically assessed. These risks can then be either eliminated or controlled at the appropriate points in the facility's life, ranging from initial design through construction, commissioning, operation, and abandonment of the facility. The SMS must be comprehensive, integrated, and contain feedback loops that continually measure performance and drive change.

The components of an SMS have been defined by the UK Ministry of Defence (MOD, 1996) as:

- Policy
- Organization

- Implementation
- Measuring
- Review and development

An SMS will include items such as the following:

- Safety policies and the organizational and facility safety objectives
- Organization reporting structures—roles and responsibilities
- Risk assessment and risk management
- Methods of employee involvement in risk management
- Employee selection, competence, training, and induction
- Integration of contractor and support services in risk management
- Design, construction, and commissioning procedures
- Safe operational procedures for normal and abnormal circumstances
- Systems of maintenance, inspection, and modification
- Systems of managing change to ensure safety
- Methods, systems, and procedures for ensuring the occupational health of employees
- Emergency response including controls, personnel evacuation, escape, and rescue
- Incident investigation and reporting, corrective and follow-up action
- The method of performance review and audit including review in the light of external experience

The SMS should ensure that all necessary linkages between system elements are identified and, where appropriate, should draw on the principles of quality management.

Auditor/assessor responsibility

All management systems must be audited on a regular basis by an independent audit team.

Auditors fall into one of three types. The first is someone from within the immediate organization who is charged with checking the quality of the safety case program. The second type of auditor works for the company (or duty-holder) that owns the facility but is in a separate (often corporate) organization. This type of auditor may also be a company hired by the facility management to mimic a regulatory audit. The third type of auditor is a government agency or other regulatory authority.

With respect to safety cases, the auditor or assessor has three key roles:

(1) Provide guidance to the owner as to what is required in the safety case.

(2) Formally accept (or reject) the safety case after it has been prepared and presented by the operator. Not only must the safety case as written be accepted, but

also the operator has to demonstrate that his organization has the ability, management commitment, and resources to implement the safety case's requirements.

(3) Ensure that the operator is actually doing what he said he would do in the safety case once operations commence. Such reviews should occur on a regular basis. The UK HSE (Health and Safety Executive), for example, requires that, "the duty holder must carry out a 'thorough review' of the current safety case at least every 5 years or as directed by HSE."

The active participation of the regulator in this manner differs from other standards, such as OSHA's process safety management program. In these cases, the regulator does not check or validate the program; it merely requires that a program exists. Only if there is an accident, is the program scrutinized by the regulator.

Program steps

The development and application of a safety case consists of the following six steps:

(1) Develop and implement an SMS.

(2) Ensure that the SMS meets the pertinent health and safety rules and regulations.

(3) Identify all hazards with the potential to cause a major accident and show how the associated risk has been determined.

(4) Show that measures are in place to prevent a major accident so the associated risks have been reduced to as level that is as low as reasonably practicable (ALARP).

(5) Show that mitigation measures are in place should a major incident occur.

(6) Ensure that adequate provisions have been made for auditing the program, and for audit reporting.

A safety case does not have to be quantified. Indeed, only half of the European countries stipulate the use of quantitative risk analysis (QRA) in their safety case legislation (Rasche, 2001). Nevertheless, it is difficult to demonstrate that risks have been reduced to the target ALARP threshold without some type of quantification.

Timing

As discussed in the previous chapter, a company that is developing a safety case from the ground up probably needs around five years to complete the work.

The following regulatory thresholds have been provided for offshore operations in the United Kingdom by the HSE (Health and Safety Executive):

- Notification to the HSE 3 months before the design is finalized.
- Submit the operations safety case 6 months before the operations are started.
- Review the operations safety case every 5 years.
- Submit the dismantlement safety case 3 months before dismantlement begins.

ALARP demonstration

Some safety case reports contain an "ALARP Demonstration" section that shows how the identified major hazards have been managed so that the associated level of risk is acceptable. ALARP and acceptable risk are discussed in Chapter 1.

LENGTH OF THE SAFETY CASE

Safety cases are usually lengthy documents, supported by detailed analyses, studies, and reports. This does not mean, however, that the safety case report delivered to the regulator and public needs to be all that lengthy. It can provide a summary of the key results and programs, and then show where additional information can be found. If the safety case is too lengthy, it may obscure the key elements of the SMS.

The following representative document sizes as seen in Table 6.1 were published by Kinsella (2010) for a medium-sized North Sea platform.

Table 6.1 Representative Safety Case Sizes

Part	Title	Pages
1	Executive Summary and Introduction	40–50
2	Description of Facilities	100–150
3	Safety Management System Description	80–100
4	Formal Safety Assessment	150–180
5	ALARP Demonstration	40–60
6	Emergency Response Arrangements	40–60
7	Performance Monitoring	40–60

MAJOR ACCIDENTS

Although a safety case can and should cover all aspects of safety (occupational, process, and technical, as described in Chapter 1), the focus tends to be on identifying and avoiding what are known as Major Accident Events (MAEs), i.e. catastrophic events such as fires, explosions, and the release of toxic chemicals. Associated with MAEs are Safety Critical Elements (SCEs) and Performance Standards.

FSAs (as discussed in the next chapter) will analyze MAEs to assess the level of risk associated with them, and to take corrective actions as necessary to reduce the risk to an acceptable level.

Major accident event

An MAE has the potential to cause multiple fatalities of persons at or near a facility. Natural events such as a hurricane can be counted as MAEs. The UK Health and

Safety Executive states that an offshore MAE generally involves one or more of the following:

- A fire, explosion, or the release of a dangerous substance, involving death or serious personal injury to persons on the installation or engaged in an activity on or in connection with it.
- Any event involving major damage to the structure of the installation or plant affixed thereto or any loss in the stability of the installation.
- The collision of a helicopter with the installation.
- The failure of life-support systems for diving operations in connection with the installation, the detachment of a diving bell used for such operations or the trapping of a diver in a diving bell or other subsea chamber used for such operations.
- Any other event arising from a work activity involving death or serious personal injury to five or more persons on the installation or engaged in an activity in connection with it.

MAEs are generally identified during the HAZIDs and other early safety analyses.

Safety critical elements

Associated with each MAE are one or more SCEs that prevent, control, or mitigate the MAE (NOPSA 2009).

> *Essential for the integrity of any installation are the safety-critical elements (SCEs). These are the parts of an installation and its plant (including computer programs) whose purpose is to prevent, control or mitigate major accident hazards (MAHs) and the failure of which could cause or contribute substantially to a major accident.*

The SCEs should reduce the risk associated with an MAE to the facility's ALARP level.

SCEs are generally identified during the hazard analysis sessions (HAZID and HAZOP) and bow-tie analyses. Their effectiveness is evaluated through a range of quantitative risk analysis (QRA) techniques.

Performance standards

Appropriate performance standards must be defined for the operation of the safety critical aspects. A performance standard is a standard, established by the operator, of the performance required of a system, item of equipment, person, or procedure, which is used as a basis for managing the risk of a major accident event.

IADC HSE CASE GUIDELINES

Although the philosophy that lies behind safety cases is that a company or facility will develop a document designed to meet its own particular needs and requirements, most facilities will have programs that are quite similar to one another. Therefore, the

International Association of Drilling Contractors (IADC) has developed a Health Safety and Environment Case Guideline for Drilling Contractors. This standard was first issued in 2003. The latest edition (IADC, 2006) is dated October 2006. The document is detailed and lengthy (150 pages).

In the introduction to the guideline, the IADC states the following:

Although the Guideline is not compulsory, its use is recommended as a demonstration of good industry practice for Drilling Contractors.

The safety case structure proposed by the IADC is as follows:

- Part 1 HSE Case Introduction
- Part 2 Drilling Contractor's Management System—HSE Management
- Part 3 MODU/Rig Description and Supporting Information
- Part 4 Risk Management
- Part 5 Emergency Response
- Part 6 Performance Monitoring
- Appendix 1 Reference Documents
- Appendix 2 Abbreviations, Acronyms and Definitions
- Appendix 3 Drawings and Schematics
- Appendix 4 Coastal Area Regulatory and ISM Code Index Indexes

The IADC structure is directed toward MODUs (Mobile Offshore Drilling Units, including jack-ups, semisubmersibles, and submersibles). However, this document can be used to help develop a safety case for a much wider range of technologies. The IADC structure is summarized in Figure 6.1, which is derived from the document's Figure 1.1.1—*HSE Management Assurance*. (The word "Part" in Figure 6.1 refers to the sections of the IADC document.)

Figure 6.1, which is very much in line with the description of a safety case system already provided shows how the Management System, Technical Information, Emergency Response, and Performance Monitoring are fed into, and are also modified by, the overall Risk Management program.

The IADC Guidelines provide very useful detailed information. For example, picking one topic at random, Section 3.5.5 of their guidelines states the following about Passive Fire Protection, shown in Table 6.2.

STRUCTURE OF A SAFETY CASE

In spite of the differences that exist between industries and facilities, it is useful to have a template as a starting point. The IADC system has already been described; another representative structure is shown below.

- Section I Executive Summary
- Section II Introduction

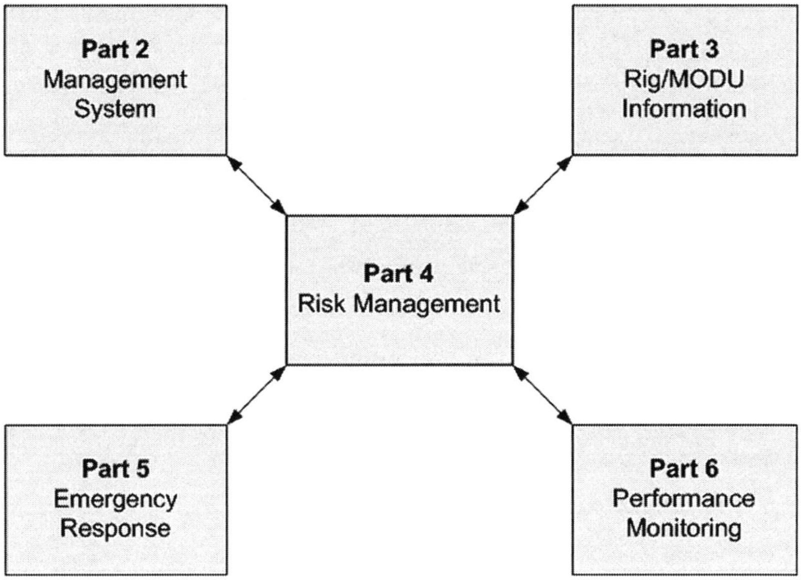

FIGURE 6.1

IADCHSE Management Guideline structure.

Table 6.2 Sample of IADC Guidance—Passive Fire Protection
Objective To describe the passive fire protection systems.
What should be (referenced) in the HSE case? Description of the passive fire protection systems including: ■ Drawing showing location and rating of fire resistant bulkheads, decks, and deckheads ■ Details of any structural fire protection for load-bearing structures ■ Details of any unprotected nonloading bearing structures and decks, which have a role in controlling fires ■ Details of any HSE critical equipment/systems that has passive fire protection ■ The use of nonflammable materials in accommodation and other occupied areas

- Section III Policies, Objectives, Regulations, and Standards
- Section IV Facility Description
- Section V Safety Management System
- Section VI Formal Safety Assessment
- Section VII Audit and Review

A discussion of each of the above sections is provided below.

Section I—Executive summary

This section should provide the reader with an overview of why the safety case was developed, the facilities and operations that it covers and who it was written for. The summary should provide a brief statement as to the assumptions, conclusions, and recommendations that were made. It is particularly important that this section highlights any major safety issues that require high priority attention.

This is the most important part of the safety case.

Section II—Introduction

The introduction to the safety case summarizes issues such as the physical scope of work, the methods used to analyze risk, the project phases that are covered, and other major factors such as the regulations that have to be addressed.

Section III—Policies, objectives, regulations, and standards

In this section of the safety case, the company management outlines its goals, and the parameters within which it is working.

Policies

The company policies regarding safety and environmental compliance are described in this section. For example, most large companies have their own standards for managing safety and engineering design. This section describes those policies and standards and shows how they are to be applied to the safety case regime, and how they align with the pertinent regulations.

Objectives

Objectives for the facility can include targets for the number of safety events, system reliability, and maintenance costs. Such objectives can affect the structure of the safety case.

More broadly, the objectives for the safety case can include the development of an enhanced safety culture in the facility, and the active participation of employees in the maintenance of a safe place of work. The safety case will need to clearly identify the methods used to involve employees in safety management.

Regulations

The use of the safety case approach does not mean that regulations, codes, and standards can be ignored. A regulation is a legal requirement—no matter how sophisticated a risk analysis may be, the regulatory requirements must be met. Therefore, the safety case should summarize those rules, show how they are being complied with, and how they are integrated into the overall safety case management system.

Standards

Although a safety case is fundamentally nonprescriptive and performance based, this does not mean that it will not refer to industrial standards and codes—many of which are quite detailed. Not only will API standards, particularly the 14 "family," be referred to, but also documents from other professional bodies such as the American Society of

Mechanical Engineers (ASME) and the Institute of Electrical and Electronics Engineers (IEEE) are likely to be cited.

Section IV—Facility description

This section provides the information needed to verify that the design and operating philosophies are consistent with the SMS and the assumptions and outputs of the FSA. In addition to a description of the facility itself, the FD should also discuss interactions with existing and planned facilities.

The FD need not contain detailed procedures, calculations, drawings, or plans, but should contain sufficient information to allow the regulator to assess whether the systems and conclusions presented in the safety case are reasonable. General documentary evidence that supports the conclusions reached in the safety case should be referenced. An assessor or regulator should be given access to the relevant documentation as necessary.

For offshore rigs and platforms, the safety case will generally contain the following minimum information:

- Ownership and location
- Operations and facilities description
- Offshore environment
- Primary systems and functions
- Safety features and emergency systems
- An overview of the facility, highlighting key assumptions and phases of development and any unique features
- A summary of key design parameters with cross-references to key technical documents (covering storm/wave/current conditions, wind, seawater/air temperatures, earthquakes, cyclones/hurricanes, other extreme conditions, and seabed stability)
- A description of the process flow and operations
- Equipment layout for all decks
- A description of the functions of the facility with reference to key processes, wellhead and utility systems, drilling, workover, wireline systems, and marine and helicopter operations
- A summary of hazardous substances that are used or stored at the facility, along with an estimate of the inventory of these substances
- A description of the design safety philosophy, features, and systems provided on the installation

Section V—Safety management system

The SMS is at the heart of a safety case. It describes the management system whereby hazards are identified and risks are systematically assessed throughout the life of the facility.

Section VI—Formal safety assessment

Once the facility description is finalized and guidance for allowable risk is provided, an FSA can commence. (The term Safety in Design is sometimes used to express the same concept.)

An FSA requires the identification and evaluation of hazards over the life of the project from the initial feasibility study through the concept design stage, to construction and commissioning, then to operation, decommissioning, and abandonment of the facility. The FSA provides the technical basis for demonstrating that a design or operation is safe. The purpose of an FSA is to identify major hazards, assess the associated risks, and to show that those risks have been reduced to an acceptable level over the life of the installation. (Hence the FSA needs to provide criteria for acceptable risk.)

The FSA should thoroughly analyze potential MAEs that could lead to fatalities or major equipment and production loss. The analysis should show what safeguards have been put in place, and how the facility would respond to a major incident.

Analyses and studies that are frequently carried out during an FSA include the following:

- Flare and Radiation Analyses
- Escape, Evacuation, and Rescue Analysis (EERA)
- Temporary Refuge Analysis
- Human Factors Engineering
- Thermal Plume Analysis
- Reliability, Maintainability, and Risk

An overview of these topics is provided in the next chapter.

Section VII—Audit and review

The importance of audits and how they are to be carried out has been discussed in Chapter 4. The audits of safety cases follow the same general principles, although the word "assessor" seems to be used in preference to "auditor." Either way, they are doing the same thing—they check that the safety case *as implemented* meets its expectations.

Audits

Factors considered in the assessor's reviews include the following:

- The operator's incident/accident experience and causal factors, complaints, legislative compliance reviews, and the operator's internal audit results
- The combined national experience of operators
- National and international trends and experience
- General industry experience and developing standards
- The effectiveness with which the commitments in the safety case are being implemented

- Monitoring the effectiveness of SMS and operator audits of them
- The degree to which the work force is involved in implementing the safety case regime

Overall, the assessor's job is to ensure that management systems are in place, that they are effective, and they are being followed. Rather than checking the details of the safety program, the assessor will evaluate management systems and their effectiveness.

Assessors will be involved in both onsite appraisal of the delivery of improvements and assessing the complex technical arguments put forward for alternative approaches. They must be able to review and evaluate the quality and effectiveness of the safety case without duplicating the work.

Performance measurement

Performance standards are the key to an effective safety system. They specify what has to be done, when, by whom and to what extent and ensure that the system is operating as planned in the achievement of objectives through linking roles and responsibilities to actions in a measurable way.

Measurement of performance has traditionally been focused on lagging indicators, such as Lost Time Injury Frequency Rates. Leading indicators such as the number and quality of safety audits conducted, the measurement of management commitment to safety through employee perception studies, and the quality of the facility safety plan can provide a real-time measure of the effectiveness of the safety management arrangements. They measure management's commitment to safety and can uncover weaknesses before they develop into fully-fledged problems.

Section VIII—References

Although a safety case will generally be large and comprehensive, not all the information used will be included in the document itself. Reference documents can be stored separately. These references will generally be of three types:

(1) Supporting documentation and calculations for detailed items in the safety case such as blast and dispersion analyses

(2) External references such as regulations and guidance from government agencies

(3) Internal references such as company codes and standards

MAINTAINING THE SAFETY CASE

Safety cases are no different from any other safety document—their value lies not in the quality of the material that is developed and presented, but in the degree to which that material is used. A short, poorly printed safety case that is dog-eared and splattered with coffee stains has achieved its goals. A thoroughly researched, carefully printed safety case that is in pristine condition may have helped in the development of a safe initial design, but it is not making an ongoing contribution to safety and efficiency.

Moreover, a safety case is out of date almost as soon as it is published. Operating conditions change, new organizations are introduced, and equipment is modified. After a year or so, the safety case is likely to be describing a facility that is substantially different from the one that is actually operating. Such a disparity is serious; indeed, having out-of-date and inaccurate information is probably worse than having no information at all. On a facility without any type of safety case, the operations personnel will accept that risk situations exist. On a facility for which a safety case was prepared, however, the operations personnel may feel that high risk situations have been addressed, and they have nothing to worry about.

Changes in risk perception

The original safety case is usually developed during the design phase of a project. It can be as much as five years from that time to the period when the facility is actually in operation. During that time, judgments regarding levels of acceptable risk and of "reasonably practicable" technology may change. After a facility has been in operation for a few years, the public perception as to what constitutes acceptable risk may shift—usually in a direction that is more conservative. For example, following the Deepwater Horizon incident BOEMRE implemented a ban on all deepwater drilling in the GoM. In spite of the controversy that this ban generated, it did show the shifting of the public's perception regarding oil spill risks.

Changes in technology also lead to shifts in perception as to what can be done to control risk. For example, the use of sophisticated, high-integrity safety shutdown systems has become more commonplace as instrumentation technology has developed and costs have decreased. Such technological and economic shifts need to be reflected in updates to the safety case.

Another potential change is regarding regulations. For example, the allowable level of a toxic gas such as hydrogen sulfide (H_2S) in the open air may be reduced.

Changes in operating conditions

Not only is there likely to be a gap between the original design assumptions and the actual operation, but the operations themselves are also likely to change. In offshore oil and gas operations, the most obvious change is that the output from the field will gradually decline and that the ratios between oil, condensate, gas, water, and solids are likely to shift. Such changes may require that the conclusions of the safety case be updated.

Bridging documents

The need for bridging documents for SEMS compliance has been discussed in Chapter 5. When two or more organizations have safety cases that interface with one another, a bridging document is also needed. For example, a floating production platform may have a floating drilling rig connected to it. Each facility has its own safety case, so a bridging document is needed to align them. The bridging documents need to consider problems at the interfaces such as the possibility that the anchors from one platform may interfere with the subsea equipment from the other.

Bridging documents are also used to create facility-specific versions of generic safety cases. For example, it has already been noted that the IADC has prepared a safety case template for a wide range of drilling operations (both onshore and offshore). A bridging document can be prepared to match the needs of a particular facility to a general template structure.

EFFECTIVENESS OF SAFETY CASES

The development of a safety case does not, in and of itself, improve safety. Safety cases are only as good as the commitment made to their preparation and implementation—an observation that is illustrated by the crash of a Royal Air Force Nimrod airplane in the year 2006 in which 14 crew members died.

A safety case had been prepared for the Nimrod. It turned out, however, that the quality of that safety case was gravely inadequate, leading to the following statements:

> *[...]the Nimrod safety case was a lamentable job from start to finish. It was riddled with errors*
>
> *[...]Its production is a story of incompetence, complacency, and cynicism.*
>
> *The Nimrod Safety Case process was fatally undermined by a general malaise: a widespread assumption by those involved that the Nimrod was "safe anyway" (because it had successfully flown for 30 years) and the task of drawing up the Safety Case became essentially a paperwork and "tickbox" exercise.*
>
> *(Haddon-Cave, 2009)*

Comments such as these emphasize that simply preparing a safety case does not *ipso facto* make that facility safe. Its development requires the commitment of substantial amounts of time from key personnel—people whose services are always in demand elsewhere in the organization.

In addition, just as employee participation is the key element of process SMSs, worker involvement is crucial to the effective application of safety cases. Unfortunately, the perception among many that a safety case is a lengthy, highly technical document that can only be understood by specialists mitigates effective employee participation. For example, the Computational Fluid Dynamics technique used to model explosion overpressure uses very sophisticated mathematics. This sophistication makes communication with nonspecialists a challenge. In addition, the sheer size and complexity of a safety case may serve as a barrier to the involvement of nontechnical personnel.

REFERENCES

DOCEP (Department of Consumer and Employment Protection (Western Australia)), September 2005. Mine Safe vol. 14 (No. 3).

Heiler, K. October 2005. Is the Australian mining industry ready for a safety case regime? 31st International Conference of Safety in Mines Research Institutes.

Kinsella, Kevin. September 2010. UK Offshore Safety Cases: Lessons Learnt. SPE Seminar, Houston.

Maguire, R. 2006. Safety Cases and Safety Reports. Ashgate.

MOD (Ministry of Defence (UK)), December 1996. Safety Management Requirements for Defence Systems. Defence Standard 00:56, Issue 2.

MOD (Ministry of Defence (UK)), December 2004. Safety Management Requirements for Defence Systems Part 1. Interim Defence Standard 00:56, Issue 3.

Rasche, Tilman. October 2001. Development of a Safety Case Methodology for the Minerals Industry - A Discussion Paper. University of Queensland.

Haddon-Cave. 2009. Charles. The Nimrod Review. Her Majesty's Stationery Office.

International Association of Drilling Contractors, October 2006. Health, Safety and Environmental Case Guideline for Mobile Offshore Drilling Units. Issue 3.2.

NOPSA, 2009. National Offshore Petroleum Safety Authority (Australia). Safety Case Approach.

CHAPTER

Formal safety assessments

7

CHAPTER CONTENTS

INTRODUCTION

The previous chapter concerning safety cases referred to the importance of formal safety assessments (FSAs). An FSA is a technical assessment carried out to determine the likelihood and impact of high-consequence events and to help develop corrective actions. Most FSA work is done during the design of an offshore facility because that is the time when major changes to equipment design and layout can be made without having to carry out expensive retrofit work.

This chapter provides some more information and guidance about typical FSAs. It must be stressed that what is provided here is only an overview of the elements of an FSA. Whole books—often very large books—have been written on each of the topics that make up this chapter.

Ian Sutton: Offshore Safety Management. DOI: 10.1016/B978-1-4377-3524-6.00007-1
Copyright © 2012 Ian Sutton. Published by Elsevier Inc. All rights reserved.

The International Maritime Organization (IMO, 2002) list the following five stages in the development of an FSA:

(1) Identification of hazards (a list of all relevant accident scenarios with potential causes and outcomes)

(2) Assessment of risks

(3) Identification of control options

(4) Cost–benefit assessment for each risk control option

(5) Recommendations for decision making

Further guidance is provided in ISO 31010: Risk Management—Risk Assessment Techniques (ISO, 2009).

An FSA should:

- Show that all pertinent regulations and standards have been addressed.
- Demonstrate that all reasonably practicable steps have been taken to ensure the safety of employees in the event of an emergency and during their transit to a place of safety. It should be demonstrated in particular that the integrity of the temporary refuge (TR), escape, and evacuation routes are maintained in the case of a major accident event, and that all reasonably practicable steps have been taken to ensure the safety of employees in the event of an emergency and during transit to a place of safety.
- Both qualitative and quantitative methods of analysis can be applied to the assessment of risk.
- It should identify the technical and other control measures that are necessary to reduce that risk to a level that is as low as reasonably practicable.
- Show that performance standards have been established.
- Show that performance is measured against those performance standards as part of the company's inspection, maintenance, and safety management systems.
- Demonstrate that there is periodic review of the FSA process by which the performance standards were established and maintained, including checks that the right things are being measured.

PHILOSOPHIES

Although not part of the formal FSA structure, it is good practice to write technical and management philosophies at the start of a project. Philosophies can be written for each design or operational activity such as material handling, blowdown, and the preparation of drilling programs. A philosophy will generally consider the following issues:

- The physical scope of work and boundary conditions for the project and the FSA. For example, the work may be confined to just the facility's topsides, with subsea engineering being excluded.

Table 7.1 Technical and Process Safety Philosophy
This philosophy outlines the general requirements to prevent personal injury or loss of life and to protect the facility from fire, explosion, and process hazards.
Prevention of personal injury or loss of life, physical damage to property, and escalation of the environmental impact shall be the primary consideration in the design and construction of the project.
Design and construction of safety inherent features and protective measures shall be provided for all phases of operation.
The primary philosophy is to follow the principles of inherent safety. This implies a systematic effort to apply the principles of "hazard elimination," "minimization/intensification," "hazard substitution," "moderation/attenuation," and "simplification." However, additional controls will still be required to control a hazardous situation, prevent escalation, and mitigate the risk to people, environment, asset, and reputation. Preferably, these safeguards will be passive or active engineered controls rather than administrative controls (i.e. dependent on direct human intervention).
Personnel are to be trained to manage operational activities with the highest regards for safe procedures and to react properly in the event of emergencies. To ensure safety of the facilities, personnel will properly inspect, maintain, and test-operate periodically the equipment with strict compliance to operation and safety procedures. Operation and safety procedures will be developed and continuously improved based on experience to help prevent and/or eliminate hazardous conditions.
Each principal area or system will be addressed by using the government rules and regulations, applicable rules and standards, American codes and standards, and API-recommended practices. Where several controlling documents appear to apply, a risk-based approach will be adopted to ensure that an equivalent level of safety is achieved.

- The standards and codes to be followed. They can be internal to the company or from an outside organization such as the API. The philosophy should also identify pertinent rules and regulations from bodies such as Bureau of Ocean Energy Management, Regulation, and Enforcement (BOEMRE), the Coast Guard, and the Environmental Protection Agency (EPA).
- The modeling techniques to be used.

Table 7.1 provides an example of a high-level technical and process safety philosophy.

Other philosophies, such as that to do with material handling, are likely to be much longer than what is shown in Table 7.1.

A philosophy will not generally contain detailed specification information such as the required width of access routes or the range of sea states; that type of information can be found either in the assumptions register or in detailed design procedures.

ELEMENTS OF AN FSA

An FSA is built around a suite of technical analyses of a wide range of topics. Some of those topics are listed below and then discussed in greater detail. Other items, such as the facility's emergency shutdown procedure, health analysis, and the blowdown philosophy, could be added to the above list.

- Assumptions register
- Hazards register
- Quantitative risk assessment (QRA)
- Facility layout and equipment arrangement
- Flare and radiation
- Material handling risk assessment/dropped objects
- Transportation
- Fire and gas detection
- Gas dispersion analysis
- Fire and explosion analysis
- Escape, evacuation, and rescue
- Non-hydrocarbon chemicals
- Greenhouse gas emissions
- Noise and vibration
- Human factors engineering (HFE)
- Reliability, availability, and maintainability (RAM)

ASSUMPTIONS REGISTER

The assumptions register contains a list of the assumptions used to develop the FSA and its matching safety case. The design philosophies already discussed will contain many of the assumptions used on projects.

The justification for the assumptions made should be provided. Generally, the justification will come from one of three sources:

(1) A public report such as the *Offshore Hydrocarbon Release Statistics and Analysis, 2002* (HSE, 2002) that provides information on leak and ignition frequencies.

(2) An industry database such as the *Offshore Reliability Data Handbook* (SINTEF, 2009) or *The Update of Loss of Containment Data for Offshore Pipelines* (HSE UK, 2003).

(3) The company's own internal sources of information and statistics.

Some of the topics that are typically found in the assumptions register are listed below.

Deck type

The results of blast and gas dispersion analyses vary significantly depending on whether the deck is plate or grate. Therefore, the assumptions made as to the type of deck to be used in various parts of the platform need to be made explicit.

Numbers of personnel and their locations

The assumptions register should specify how many people are on the platform, and where they are most likely to be located. An estimate as to peak manning loads, for example during drilling, should also be provided.

Leak size and frequencies

The assumptions made as to the size of leaks from flanges, fittings, piping, instruments, and vessels need to be documented. Similarly, an estimate as to the frequency with which leaks can occur is required. The frequency value will generally vary inversely with hole size.

Transportation logistics

An estimate regarding the number of journeys to be made by helicopters, workboats, and other forms of transportation is required. The consequence of accidents should be estimated. Assumptions to do with ship collisions (including pleasure boats that may be present) should be made clear. Factors to consider include the speeds at which collisions may occur, and whether collisions occur while boats are maneuvering or drifting.

Lifting operations

Assumptions to do with lifting operations need to be spelled out as a basis for the material handling study. Issues to be itemized include the following:

- The types of lifting devices (monorails, platform cranes, and chain hoists)
- The use of cranes on service boats (if available)
- Areas for potential dropped objects (including subsea)
- Loading and unloading supply boats

Guidance should be provided as to the percentage of drops that occur over the deck, over the side (into the sea), and onto a workboat.

Rescue and recovery operations

Assumptions as to the effectiveness of emergency response and rescue operations need to be spelled out.

Meteorological ocean data

The assumptions register should contain meteorological information, covering both normal and extreme weather conditions. The information should include the following:

- Mean wind speed
- Stability class

Table 7.2 Representative Structural Failure Times

	Failure Time (minutes)	
	Jet Fire	**Pool Fire**
H-60 rated firewall	10	60
J-15 rated firewall	15	-
Steel beam	5	10
Steel plate	5	10
Pipe/riser/process vessel	5	10
Riser SSIV	7	10
Jacket leg	15	30

- Sea conditions and loop currents
- Mean air temperature
- Mean humidity

Structural failure time

Assumptions have to be made regarding the time it takes for steel structures to fail when they are exposed to fire. An example is provided in Table 7.2.

HAZARDS REGISTER

The documentation of hazards analyses has been discussed in the chapters covering SEMP and SEMS. Some companies elect to include the hazards register with the FSA. This register will certainly have to be provided if the FSA is part of a safety case.

QUANTITATIVE RISK ASSESSMENT

One of the principal differences between the SEMS and safety case approaches to managing offshore risk concerns the use of QRA, also called probabilistic risk assessment or analysis (PRA). QRA is commonly included in safety case analyses (although it is not a requirement to do so). SEMS does not require a quantification of risk.

The publication of the Reactor Safety Study (NRC, 1975)—often referred to as the Rasmussen Report after the name of principal author, or by its subtitle WASH 1400—demonstrated the use of probabilistic methods in what was then the fledgling nuclear power industry. Consequently, this industry was an early developer and user of techniques such as fault tree analysis (FTA) and event tree analysis (ETA). Also, because nuclear power plants are technically quite similar to one another, the industry has also been able to set up reliability databases to which most facilities contribute.

It will be recalled from Chapter 1 that risk can be defined as shown in Eqn 7.1.

$$\text{Risk}_{\text{Hazard}} = \text{Consequence} \times \text{Predicted Frequency}............ (7.1)$$

A PRA aims to quantify Eqn (7.1). The identification of hazards will always be qualitative, but numerical values can be assigned to the consequence and frequency terms.

For example, the hazard of overfilling an onboard methanol tank could result in a fire leading to a fatality. If this event occurs, say, once every 5000 years, then the risk associated with overflowing a methanol tank is 0.0002 fatalities yr^{-1}. If the risk management philosophy requires that a fatality rate of more than one in 10,000 years is unacceptable (the ALARP value discussed on page 20), then the methanol overflow hazard needs to be ameliorated. If the hazard cannot be removed altogether, then the consequence term can be decreased, say, by reducing the size of the spill so that the resulting fire does not result in fatal injuries, or the predicted frequency can be lowered, for example, by putting in additional safety instrumentation.

Importance ranking

In the late nineteenth century, the Italian economist and misanthrope Vilfredo Pareto famously noted that most of the wealth in a community was held by a small proportion of the population. From this insight, he developed the 80/20 rule, or the Pareto principle, whereby 80% of the wealth is owned by about 20% of the population. His principle, which has no theoretical underpinning, is widely observed to be true in many fields of human activity (Juran, 1951). It is applied here to the topic of risk analysis in the process industries.

The Pareto principle can be expressed mathematically as shown in Eqn 7.2.

$$\log n = c + m \log x .... (7.2)$$

where n is the number of items whose value is greater than x and c and m are constants.

Examples of the principle's applications in an industrial context include the following:

- 80% of a company's sales come from 20% of its customers.
- 80% of a company's sales are made by 20% of the sales force.
- 20% of the workers are involved in 80% of the accidents.
- 20% of the equipment items cause 80% of the facility shutdowns.
- 20% of a company's products will account for 80% of the total product defects.

In each of the above examples, an "important few" or the "vital few" have a large impact on the business, whereas the "unimportant many" are much less significant. Therefore, a safety manager should direct his or her program toward that minority of workers who create or suffer the most accidents. Spending time on the "unimportant many"—those whose behavior is generally safe—is not likely to have much benefit.

The same approach can be used in technical and process safety—it will be found that most events are caused by a relatively small number of actions. An example of this can be seen in the selection of just four of the SEMP elements as being the "most important" in the first version of SEMS.

One commonly held misconception of the Pareto principle is that 80% of the problems can be resolved with 20% of the resources. In fact, the principle makes no statement at all as to how much effort is needed to address the contributing factors.

Fault tree analysis

Risk can be analyzed in one of two basic ways: inductively or deductively, i.e. either bottom-up or top-down. In a deductive analysis a system failure is postulated. The analyst then works backward to deduce what combinations of events could have occurred for the system failure to have taken place (a detective solving a crime is thinking deductively). FTA, the topic discussed in this section, is deductive.

An inductive analysis works in the other direction. A single failure, such as a pump stopping or a valve closing at the wrong time, is postulated. The inductive approach then determines what impact the item failure could have on the overall system performance. ETA is inductive. Both techniques provide a clear and intelligible way of determining the combination of events needed for an undesirable incident to occur. Their strict logic cuts through the "I think/You think" discussions that are the bane of so much risk analysis.

First used by the Bell Telephone Laboratories and the Boeing Corporation in the years 1962–1964 to analyze potential problems with the Minuteman missile launch control system, FTA provides a clear and intelligible way of determining the combination of events needed for an undesirable incident to occur (Vesely, 1981). The graphical nature of the analysis can help managers, engineers, and operators better understand how their systems can fail. Moreover, it is often found that the rigor and logic of a FTA stimulates creative thinking, and allows experts to add their experience and opinions in a structured manner. So the fault tree method—in spite of the fact that it is based on logic and Boolean algebra—can help identify new hazards and previously unthought-of failure mechanisms.

Once a fault tree has been developed, failure rate data for individual components in the system can be entered into the tree so that an estimate of the likelihood of the undesired event (the "top event") can be made. Frequently, the quality of the failure rate data is poor; nevertheless, through use of the Pareto principle or 80/20 rule discussed above, a quantified analysis still provides useful insights because it identifies what items in the system contribute the most to system failure. Moreover, once the model has been developed, and preliminary estimates as to failure rates have been made, case studies that examine changes to the process and the effects of additional safeguards can be carried out. Also, as improved data for equipment failure rates and repair times becomes available, the quality of the analysis will improve.

FTA has a reputation for being time consuming and costly and for requiring the services of highly specialized risk analysts. This reputation came about because most of the original FTA work was carried out in the aerospace and nuclear power industries. In both of these industries, the consequences of an accident are very severe so highly detailed analysis was required. This perception that FTA is extremely time consuming may be one reason that the technique has not been more widely adopted by the other industries. (Another reason for the method's lack of general adoption may be that fault tree diagrams are not immediately intuitive; the managers who are the ultimate customers of the analysis require some training in how to understand the results.)

In fact, fault trees do not have to be large, nor does their development have to be excessively time consuming. Even a small, quickly-executed fault tree can provide useful and unique insights into how a system can fail, and what can be done to reduce the associated risk. It is quite feasible to develop a useful fault tree inside the space of a few hours. Moreover, sophisticated software is not needed for smaller FTAs. Simple trees can be drawn using basic graphics packages, and their quantification can be executed with spreadsheets or even hand calculations.

Some of the technical terms used in fault tree and ETA are explained below. In this book, words such as "or," "and," and "if" are put in small capitals when referring to the elements of a fault tree. Doing so helps distinguish the use of those words when used in normal discursive text.

Gates

A fault tree is built up of gates and events. The principal types of gate are as follows:

- OR Gates
- AND Gates

Although other types of gates are sometimes used, they can all be created from a combination of OR and AND Gates.

OR Gate

An OR Gate is a logic gate that gives a positive output if one or more of the inputs to the gate are positive. The symbol for an OR Gate is shown in Figure 7.1. Various labeling conventions are used in FTA. The system used in this book is to put a cross inside the OR Gate symbol. Other analysts use a plus sign to convey the same meaning.

An example of an OR Gate is the source of ignition for a fire hazard. For example, four possible sources of ignition may exist:

- A naked flame (say from a welding torch); OR
- A spark from electrical equipment; OR
- A hot surface (such as the outside of a fired heater); OR
- Lightning.

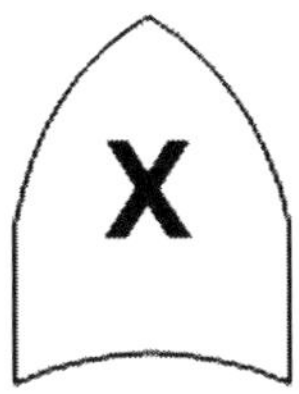

FIGURE 7.1

OR Gate.

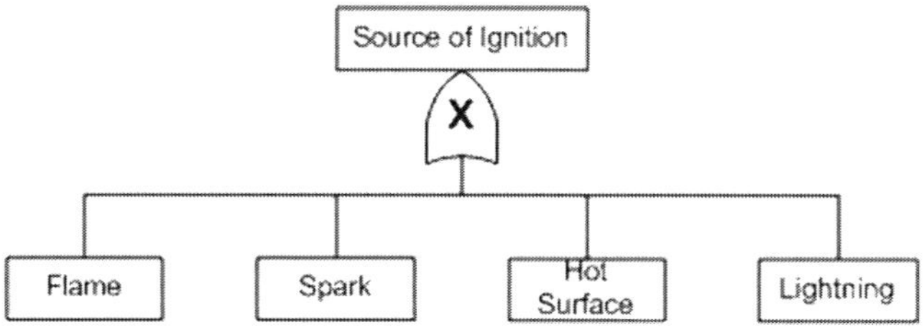

FIGURE 7.2

Example of the Use of an OR GATE.

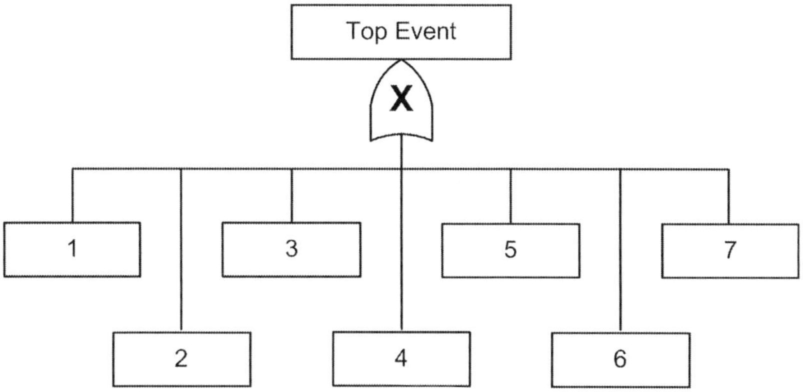

FIGURE 7.3

OR GATE with Seven Inputs.

The OR Gate for these four items can be written in the following words:

> *IF a naked flame exists OR IF a spark exists OR IF a hot surface is present OR IF lightning is present, THEN a source of ignition exists.*

Figure 7.2 shows the OR Gate that corresponds to the above logic.

There is no theoretical limit to the number of events that can enter an OR GATE. However, more than five inputs can be difficult for people to visualize and understand. In such circumstances, it generally makes sense to create two or more OR Gates; the output from each then enters a higher level OR Gate as illustrated in the transition from Figure 7.3 to Figure 7.4.

The logic and mathematics to do with the system analysis is not changed by dividing the gate into two; the division is made just to provide readers with a better understanding as to what is going on.

AND Gate

The second type of logical gate used in fault trees is the AND Gate, the symbol for which is shown in Figure 7.5. All the inputs to an AND Gate have to be positive for the output to be positive.

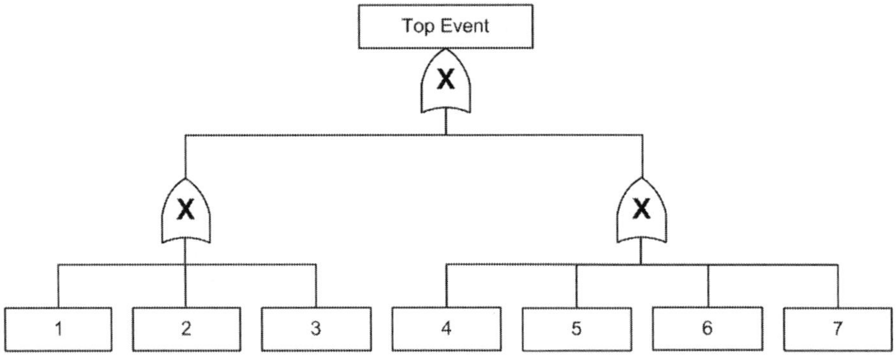

FIGURE 7.4

OR Gate Divided into Two.

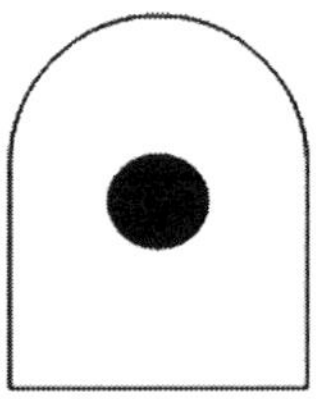

FIGURE 7.5

AND Gate.

AND Gates are frequently found in safety systems. One of the events entering the AND Gates represents an item or equipment failure; the other input represents a safety system whose purpose is to protect the system against the consequences of that failure. For this reason, the more AND Gates that can be introduced into a fault tree, the more reliable and safer the system will be.

As with the OR Gate, there is no mathematical limit to the number of inputs to an AND Gate. However, it may also make sense to divide the gate into subgates so that the tree is easier to understand.

The classic "fire triangle" can be visualized as an AND Gate. For a fire to exist, three conditions must be met:

- Fuel above its flash point must be present.
- Oxygen (usually as a component of air) must be present.
- An ignition source such as an open flame or hot surface must exist.

The above logic can be expressed as follows:

IF fuel is present AND IF oxygen is present AND IF an ignition source exists, THEN a fire will occur.

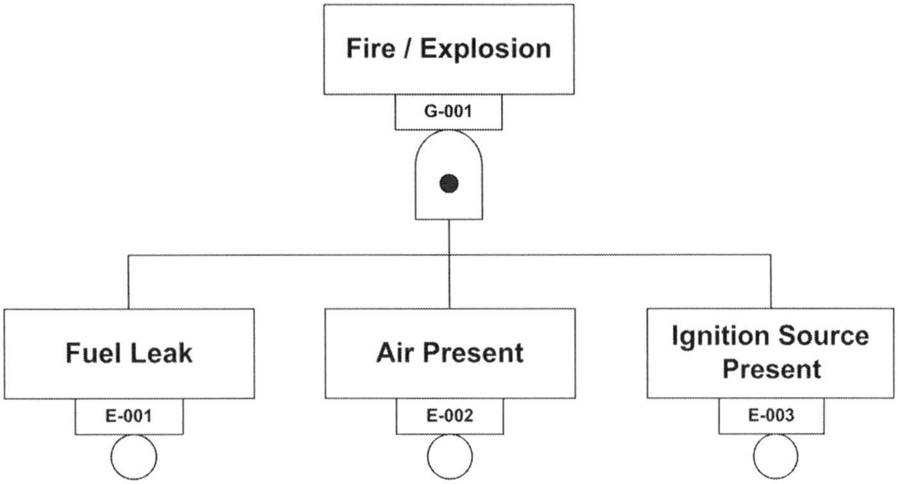

FIGURE 7.6

Fire Triangle.

The fault tree gate corresponding to the above statement is shown in Figure 7.6.

The events and gates in Figure 7.6 each have a description and a label associated with them. The description is intended for use by persons reading the results of the analysis. It provides detail as to the purpose and function of that event or gate. The label is a unique identifier used by the fault tree software. Generally, the label will be composed of a letter and a number. The letter "G" indicates a gate; the letter "E" an event (any type).

It is important to give unique labels to each of the gates and events for the following reasons:

- Different events may have the same name. If the process contains two tanks, then the event "tank overflow" can occur in two places. However, each will have its own distinct label.
- Some events occur in more than one place on the tree. In the worked example, the event "instrument plugs"—E-004—occurs three times. This is a common cause effect. Common cause events need to be identified properly before the tree is quantified.

Figure 7.6 can be further expanded, as shown in Figure 7.7, by using an OR Gate to expand the "ignition source present" term.

Event tree analysis

ETA uses the same logical and mathematical techniques as FTA. However, whereas a fault tree analyzes how an undesirable top event may occur, an event tree considers the impact of the failure of a particular component or item in the system, and works out

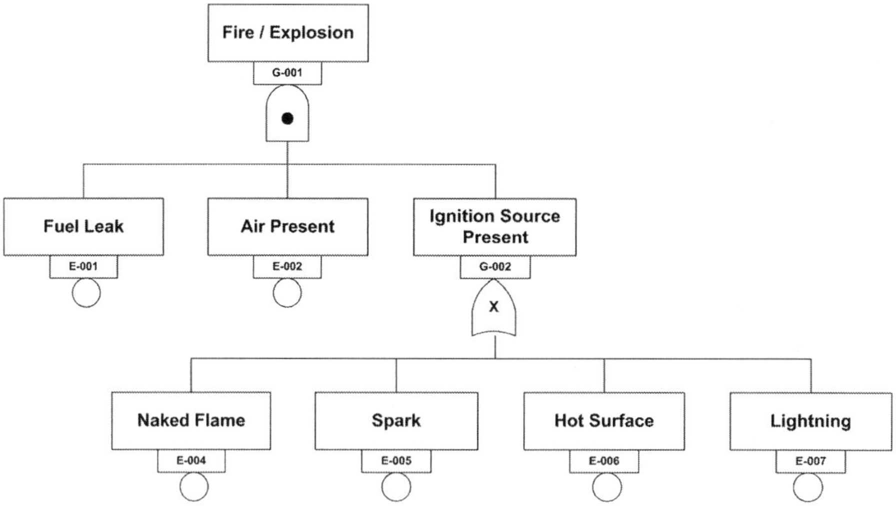

FIGURE 7.7

Expansion of the Fire Triangle.

the effect such a failure will have on the overall system risk or reliability. Event trees are inductive; fault trees are deductive.

The initiating event in an event tree will usually fall into one of the following four categories:

(1) Failures or unsafe conditions in individual items of equipment

(2) Human error

(3) Utility failures

(4) External events (such as hurricanes or earthquakes)

Figure 7.8 shows an event tree for a situation in which the pressure in a vessel rises. It starts with the event "high pressure in the vessel." The reasons for the occurrence of this event are not explained. There are four layers of protection, each of which has a chance of either success or failure. At the first junction, the normal control system acts. If it brings the pressure to a safe state, then there is no need to continue with the development of the tree. Therefore, the "success" lines are dashed, showing that they do not need to be considered.

The "failure" lines in Figure 7.8 continue through three more layers of protection. If all the protective devices and systems fail, the tree follows the thick line that leads to rupture of the vessel.

The use of the event tree method is most appropriate when there are many safeguards and protective layers between the initiating event and the final outcome. In the nuclear industry, the scenario of greatest interest is "loss of cooling to the reactor core." Were this event to occur and the subsequent safeguards not to work properly, then the reactor could melt down. Obviously, such a scenario is very serious, so the engineers responsible for designing nuclear power plants incorporate many layers of safety, thus

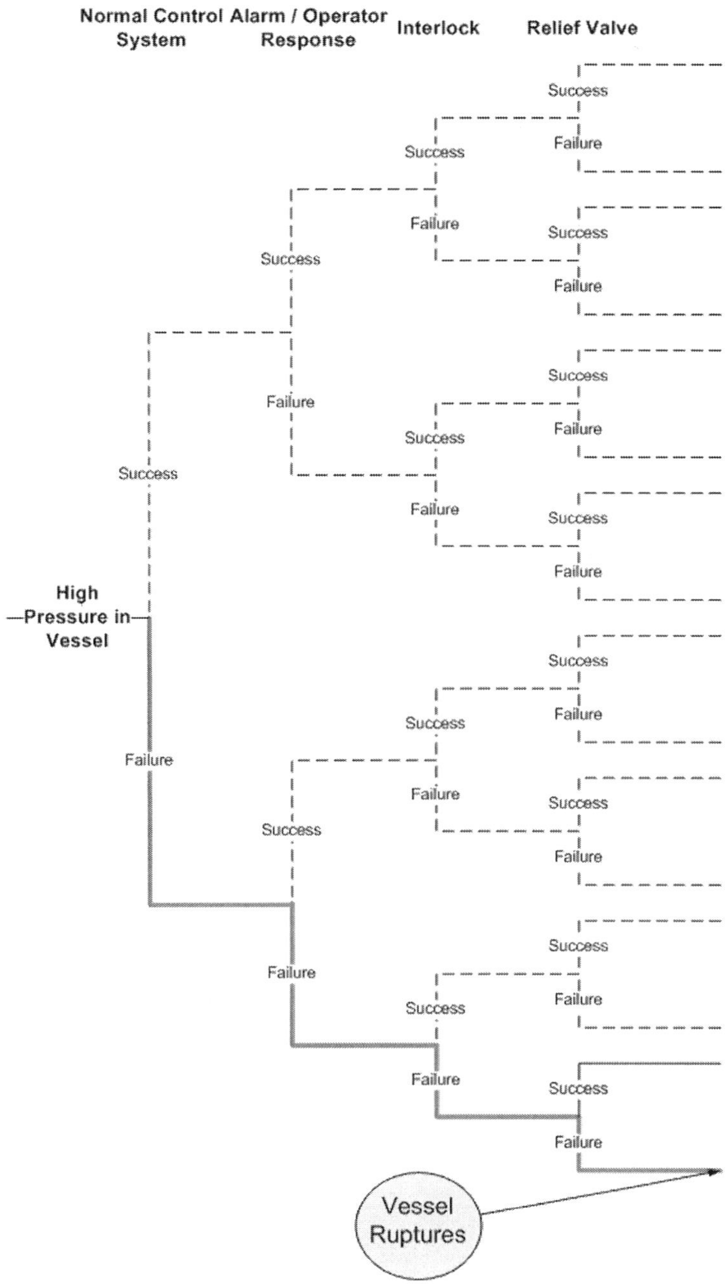

FIGURE 7.8

Event Tree Example.

making the chance of an accident remotely small. (In the case of the Fukushima Dai-ichi accident, for example, the initiating event was the earthquake. The subsequent layers of protection such as the diesel-powered cooling water pumps all failed until the final event—core meltdown—occurred.)

Offshore facilities rarely have so many layers of protection because the worst-case accidents are not usually as serious as a nuclear accident, hence the corresponding event trees are much shorter.

Combining event trees and fault trees

Fault trees and event trees can be combined. The gates in the event tree are treated as top events of multiple fault trees. So, for example, one of the event tree gates could be "loss of electrical power." That term then becomes the top event of a "loss of electrical power" fault tree. When using fault trees as subsets of event trees, it is important to identify the common cause or interdependent events and to enter them separately into the tree's structure.

Figure 7.9 shows how fault trees are linked into event trees. The likelihood for some of the lines in the event tree is calculated using the top event of a fault tree.

Event trees and fault trees can also be linked as shown in Figure 7.10, which is based on an ISO standard (ISO, 17776:2000).

In Figure 7.10, the fault tree (which is left to right rather than the normal bottom to top) generates a top event. This event is, in turn, the initiating event for the event tree that follows. For example, a series of equipment, human, and instrument failures could lead to the top event of "liquid overflow from a tank." The safeguards to mitigate the top event are shown in the event tree.

The bow tie technique uses the same structure as Figure 7.10.

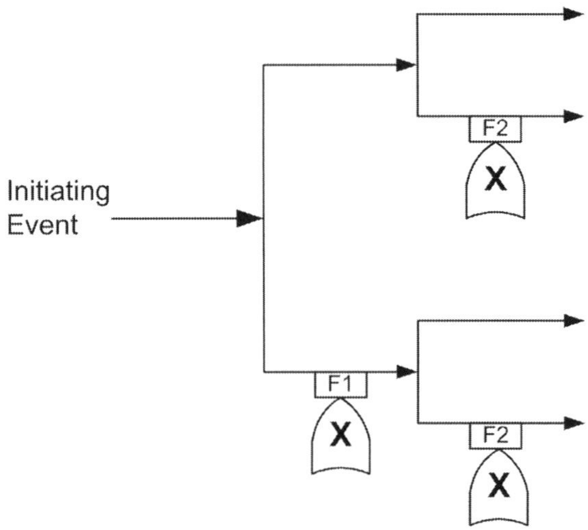

FIGURE 7.9

Events Trees and Fault Trees.

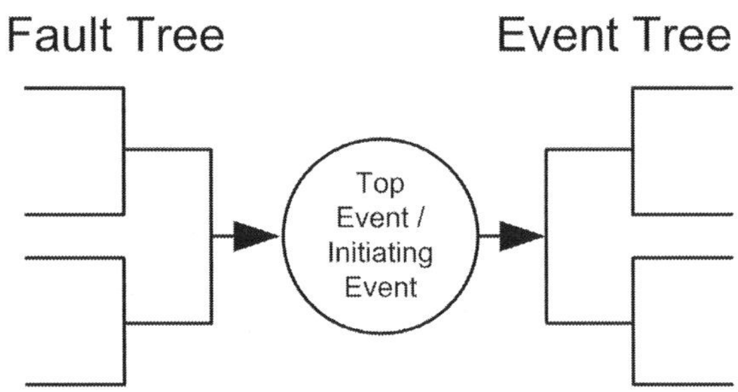

FIGURE 7.10

Events Trees and Fault Trees.

FACILITY LAYOUT AND EQUIPMENT ARRANGEMENT

Many of the safety issues to do with offshore facilities are concerned with the layout of equipment on the deck. Space is always extremely limited, which leads to problems in the following areas:

- Escape routes are difficult to design, given that there are so many equipment items and structural members that can get in the way of someone who is trying to find his or her muster station.
- Radiation from the flare can affect the safety of workers, and is also a potential source of ignition for any spills.
- It can be difficult to move objects by crane without creating the potential for dropping a heavy object onto process equipment.
- A release of toxic gas is likely to occur close to where workers are located and hence can be difficult to escape from.
- Similarly, fires and explosions are likely to occur where personnel are working, and could easily damage critical equipment and instrumentation.
- It is difficult for workers to get away from noisy equipment.

Guidance on layout, and its relationship to inherent safety, is provided by the United Kingdom Offshore Operators Association (UKOOA, 2003). Layout issues to consider include the following:

- Avoid conducting different hazardous operations at the same time.
- Separate major components containing hydrocarbons (e.g. risers, wells, and separators) from one another.
- Locate temporary quarters away from major hydrocarbon inventories, in particular wellheads and risers.
- Reduce congestion in process areas.
- Site high-pressure gas and liquefied petroleum gas (LPG) inventories in well-ventilated areas and away from large inventories.

- Locate risers to avoid supply boat impacts.

Additional items to consider include the following:

- Primary and secondary access and egress (normal operations and emergency) are clearly defined.
- Clearance is required to allow proper maintenance of equipment, to allow the proper removal of machinery or parts, and to allow entrance into equipment.
- Clearance is required for the temporary installation or removal of equipment.
- Clearance is required for lifting areas above and/or under equipment to allow easy removal without obstructions.
- There should be adequate clear space around emergency equipment such as fire extinguishers, and other safety devices such as eyewash and safety showers and air mask stations.
- Clearance is required to operate valve handwheels, sample points, and pushbutton stations.
- Clearance is required to allow installation of bolt tensioning equipment at equipment nozzles and pipe flanges.
- There should be sufficient space for operators to read manometers, thermometers, and/or readable instruments to allow proper operation by operators from dedicated locations.
- All equipment requiring regular maintenance and/or operation shall be accessible from grade or a platform.
- Layout should be based on the physical dimensions of human beings.
- Maximum manual force to be applied for any action (lifting, pushing, and pulling) should not exceed a specified value.

FLARE AND RADIATION ANALYSIS

Most offshore facilities have an emergency flare (routine flaring is generally prohibited due to environmental and climate change concerns). The radiation from a full bore flare can be intense and sufficient enough to burn workers who are in the general area. Even if the flare does not actually burn someone, the radiation can cause extreme discomfort. Moreover, the radiation can set fire to any hydrocarbons or flammable materials that may be present, and can damage or destroy equipment and instrumentation.

Table 7.3 provides guidance as to representative values for the effects of radiant heat.

For most facilities, an effective means of mitigating the effect of radiant heat is through the use of firewater, which can absorb around 9000 BTU/gallon. If the heat from a flare is unacceptable, then water can pumped up to form a curtain at the base of the flare.

Table 7.3 Radiant Heat Effects

Item	Intensity BTU hr^{-1} ft^{-2}
Solar radiation on a hot summer day	320
Continuous exposure (no evacuation required)	500
Immediate evacuation from area required	1500
Damage to exposed skin within 1 minute	1760
Damage to exposed skin within a few seconds	3000
Plastic melts	4000
Plant equipment damage	7000
Full storage tanks	10,000
Cotton clothing ignites within a few seconds	12,800
Spontaneous ignition of wood	20,000

A radiation analysis will determine what levels of radiation can be anticipated. The results are then compared with the levels of radiation that people can tolerate over various periods of time. If the results show that the radiation levels are too high, options that can be considered are as follows:

- Extend the flare boom (assuming that the facility is still in the design stage).
- Modify the flare tip.
- Provide shielding at the deck level (the shielding can be metal plates or a wire mesh).

Generally, the flare is the only source of significant radiation. However, the same approach to analyzing and mitigating other radiation sources should be followed.

MATERIAL HANDLING ASSESSMENT/DROPPED OBJECTS

One of the greatest hazards on an offshore facility is that of dropped objects, usually from one of the platform cranes. Dropped objects include drill pipes, chemical totes, galley supplies, and production risers.

Objects can be dropped either on the deck or overboard. If something lands on the deck, it could injure or kill a worker who happened to be in the vicinity. It could also seriously damage equipment or cause a release of gas or toxic materials. Objects dropped overboard, particularly in deepwater, can be traveling very quickly by the time they reach the seabed and so can do considerable damage to subsea equipment and pipelines. One study calculated terminal velocities for objects dropped in 8000 ft of water to be in the range of 1.8—3.7 m/s (4—6 miles/hour).

Before a dropped objects study is carried out, it is normal for a material handling philosophy to be developed. This philosophy will discuss the following issues:

- How objects are to be moved around the platform or rig
- How supplies and personnel are to be transferred from and to supply boats
- The location of laydown areas

Subsea dropped objects

When an object is dropped into deepwater, it does not fall straight down—it will have a cone-like drop profile. Factors that affect the sideways movement of the object include the initial crane movement, the shape, size, weight, and density of the object, stillwater motion (the lateral motion of objects due to hydrodynamic effects as they travel through water without considering the effect of currents), and subsea currents (which are variable and change at different depths). Typical values for the radius of the impact zone in deepwater are 50–100 m.

The steps in the analysis of a subsea drop are as follows:

(1) Based on a material handling study, identify the hazards associated with the movement of heavy objects, e.g. a drop from a crane transferring equipment from a supply boat to a deck laydown area.

(2) Identify drop zones based on the various lift paths.

(3) Determine the impact of a dropped object in each of the drop zones.

(4) Estimate the frequency of objects being dropped for each of the zones.

(5) Estimate the risk associated with the various material handling activities.

(6) Take corrective action as required.

The consequences associated with a subsea dropped object can be categorized, and can then be used in the hazards analysis in the follow way:

- *Low.* The lift has the potential to damage subsea equipment, including umbilicals, but will not have a significant safety or environmental impact.
- *Medium*. There is a potential for releasing a significant quantity of oil (say, 100 barrels or more) and for damage up to, say, $10 million.
- *High*. There is a potential for releasing more than 100 barrels (or whatever limit is set) and losses in excess of $10 million.

The following strategies can be used to reduce the risk associated with a dropped object to an acceptable level:

(1) Eliminate the need for the lift, if possible.

(2) Reduce the number of lifting operations. For example, rather than lifting chemical totes on to the deck, it may be possible to pump chemicals from the supply vessel to the platform.

(3) Create a safe zone on the seabed over which most of the lift will be carried out.

(4) Reduce the likelihood of a drop through the use of secondary retention devices.

(5) Reduce the severity of an event through the use of equipment such as added buoyancy or drogue chutes.

(6) Increase the separation between subsea components.

(7) Shield critical components, or place them close to larger, more robust items. In practice, studies often show that the impact energy of a dropped object can be in the region of 100 kJ, whereas an impact resistance of only 5 kJ can be achieved, so shielding is not likely to be an effective safeguard.

(8) Design subsea components so that they are less likely to fail if struck by a dropped object.

(9) Consider means of operating subsea equipment at lower than hydrostatic pressure.

(10) Add redundancy to subsea systems, where possible.

(11) Develop dropped objects emergency response plans.

Moving objects subsea

The movement of objects subsea requires a good deal of planning. Items to consider include the following:

- A hazards analysis should be carried out to identify what may go wrong, and to provide some level of risk assessment—particularly with regard to worst-case consequences.
- Adequate spare equipment should be in place before the lift starts.
- All operators and other affected personnel need to be informed as to what is going on, and should receive special training as required.
- Operating procedures should be written, as required, particularly for any lifts that are nonstandard.
- The planned work should be integrated with the management of change and simultaneous operations programs.

Safe zones

One means of reducing the risk associated with dropped objects is to create a safe zone or quadrant at the seabed. No subsea equipment is installed in that zone. All lifts to and from the rig or platform are carried out in the water column above that zone. Then, if the object is dropped, the chance of an important piece of equipment being hit is minimized.

Once the object has been lowered close to the seabed, and is still above all the subsea equipment, it can be moved horizontally to its final location. Now, should it be dropped, its terminal velocity at the seabed will be low.

TRANSPORTATION

Transportation incidents are usually to do with ships and other vessels and helicopters.

Helicopters are used to transfer personnel to and from offshore facilities. The following guidance applies to helicopter operations offshore:

- Do not use the helideck for storing any type of equipment.
- Ensure that a firefighting team is present during takeoff and landing.
- Each helideck stairway should have its own foam hose reel station and a wheeled dry chemical extinguisher.
- A drainage system should be installed around the helideck perimeter for collection of fuel spills.

The UK Civil Aviation Authority (CAA) provides guidance for the safe operation of helicopters (CAA, 2008).

FIRE AND GAS DETECTION

The best response to an emergency is to know about it as soon as possible, which means that instruments for detecting fire and/or gas releases should be provided in all sections of the facility. The signals will go to a central fire and gas detection system, which will call for the appropriate response to the alarm (either from the operators or the automatic instrumentation). Responses can include the following:

- Provide warning to the operators, who can start manual firefighting as soon as possible.
- Provide an alarm to workers in specific areas so that they can evacuate those areas.
- Start fixed fire suppression systems and special systems such as CO_2 deluge.
- Shut down heating, ventilation, and air-conditioning (HVAC) and electrical systems.
- Start fire water pumps in standby mode.
- Initiate a partial or facility-wide shutdown.

A fire or gas alarm will be communicated through audible alarms, usually supplemented by a public address system. In high-noise areas (80 dBA and greater), visible strobes can also be provided.

Fire detection

Fire detection systems used in the process industries are listed in Table 7.4, along with a summary of the advantages and disadvantages of each.

Table 7.4 Fire Detection Systems

Type	Advantages	Disadvantages	Applications
Fire eye (ultraviolet)	High speed High sensitivity Moderate cost	Potential for false alarms Blinded by thick smoke	Outdoors or indoors
Fire eye (infrared)	High speed Moderate sensitivity Easy to test manually Moderate cost	Affected by temperature Subject to false alarms from the many other sources of IR radiation	Outdoors or indoors
Smoke detectors (ionization)	Detects smoldering fires Low cost	Easily contaminated Affected by the weather	Indoor use
Smoke detectors (photoelectric)	Detects smoldering fires Low cost	Easily contaminated	Indoor use
Thermal/heat detectors	Reliable Low cost	Slow Affected by the wind	Indoor use
Rate of heat rise detectors	Self-adjust for temperature and ambient conditions Rapid detection of growing fire	Affected by the wind	Indoor use
Fusible links	Does not need electricity High reliability Low cost	Very slow Heat must impinge	Outdoors or indoors
Low oxygen detectors	Warn of accidental release of inert gas	Do not warn of fire directly	Indoors—especially in confined spaces
Combustible gas detectors	Warning occurs before the fire starts	Require more than one instrument to confirm a release	Outdoors or indoors Can be portable
Manual alarm call points	Does not rely on instrumentation	Likely to be slow False alarms possible	Outdoors and at key locations indoors

Fire eyes/flame detectors

A fire eye or flame detector detects the radiation from a flame. It requires line-of-sight capability—there must be no blockages between the instrument and the potential fire locations. A fire eye's field of vision usually covers a larger area than that of a heat detector, but it will not detect a smoldering fire as quickly as some smoke detectors.

Flame detectors are not affected by airflow characteristics. They are suitable for inside or outside use, but they must be shielded from external sources of ultraviolet or infrared radiation such as welding arcs, lightning, or radiating black bodies such as hot engines or manifolds. Ideally, the fire detection system should have more than one fire eye detecting a fire so that false alarms can be weeded out. Flame detectors can be installed inside the enclosures of all engine-driven equipment, including turbine-driven generators, compressors, and emergency and essential generators.

Older types of fire eye detectors, which worked in the ultraviolet range, sometimes had difficulty distinguishing the fire radiation and other sources of radiation, such as that from a lightning bolt. Modern detectors, many of which use infrared, generally do not suffer from this defect.

Smoke detectors

Smoke detectors are particularly useful in those situations where the fire is likely to generate a substantial amount of smoke before temperature changes are sufficient to actuate a heat detection system and before a fire eye will detect a flame. Smoke detectors use a photoelectric beam between a receiving element and light source. If smoke obscures the beam, an alarm is sounded. There are also refraction-type models that measure the light changes that occur within the instrument when smoke particles enter it.

Area smoke detectors are generally installed in buildings and accommodation areas. Where this is not practical—say in the galley area—other types of fire detector should be used. Actuation of a single smoke detector will initiate a fire alarm. If additional detectors sound an alarm, the equipment in the area of the fire and HVAC systems will be shut down.

Heat detectors

Heat-detecting devices fall into two categories: those that respond when the detection element reaches a predetermined temperature (fixed-temperature types) and those that respond to an increase in temperature at a rate greater than some predetermined value (rate-of-rise types). The two types can be combined into a single instrument. They are generally installed when the use of smoke detectors is not practical, or as a backup to smoke detectors. They are used in the following locations:

- Engine-driven equipment enclosures
- Living spaces
- Maintenance workshops and laboratories
- Machinery and pump rooms
- Electrical rooms

Actuation of a single thermal fire detector can be considered a confirmed fire condition, resulting in actuation of appropriate shutdown and fire-protection actions.

Fusible links

Fusible links are made of low-melting-point materials designed to vent pneumatic systems as the fire melts the link. The depressurization can open fire deluge valves. Fusible links are very reliable, but they do require that the fire be well under way before they work, whereas other detectors, such as fire eyes, act more quickly. Depressurization of a fusible loop is considered to be a confirmed detection of a fire, and will automatically initiate appropriate shutdowns and activate fire-protection equipment.

Low oxygen detectors

Areas that could be flooded with nitrogen, carbon dioxide, or halon-like materials should have oxygen detectors installed. Their use is particularly important in electrical switch gear rooms because inert gases are used to suppress fires. It is vital to know if the inert gas is accidentally leaking into the confined space.

Combustible gas detectors

Combustible gas detectors are generally installed in buildings and in the intakes to the HVAC air ducts. They can also be installed in outdoor areas that could have hydrocarbon vapor present, particularly remote areas such as truck unloading stations that may not have personnel present all the time. They should always be installed in the living quarters (LQ), high-value computer facilities, and offices that store vital records.

They will typically have two levels of alarm: 20% LFL and 60% LFL. If multiple detectors are installed in a single location, then a voting system can be installed. For example, if just one 20% alarm sounds, then all hot work must stop but other work can continue as normal. If three 20% alarms or one 60% alarm sounds, then an emergency response is called for.

Special types of detector will warn of the presence of hydrogen, carbon monoxide, or hydrogen sulfide.

Manual alarm call points

Sometimes, an emergency will be detected by a person rather than by the instrumentation. In such situations, manual alarm call points (MACs) are used to declare an emergency and to activate the emergency response system. MACs are generally located at entrances to buildings and at strategic positions throughout process units, including escape routes. Each call point should be accessible from at least two different locations.

The typical MAC is of the open contact "break glass" type, suitable for Division 1 locations. MACs should be covered with a guard to prevent inadvertent alarm activation. Alternatively, the MAC can be actuated by a pulling action (to prevent spurious trips caused by someone pushing the button by mistake). The emergency response system should tell the operators and emergency response team which MAC was activated.

GAS DISPERSION ANALYSIS

A leak of gas can be very hazardous. If the gas is flammable, a major fire can ensue; if the gas is toxic, workers may be overcome. There are generally three sources of gas:

(1) A release of produced gas from process equipment

(2) A release of a toxic gas from a chemical tote or storage tanks

(3) The exhaust gas that is emitted from the facility's turbine exhausts

The study of dispersed gas behavior has two major components. The first is to determine concentrations of gas downwind of the release point. These concentrations will depend on a plethora of factors such as the density of the gas, the amount released, weather conditions, and the roughness of the ground surface. The second part of the gas modeling has to do with the effect of the gas on the human body. (Some gases, notably nitrogen, simply replace the oxygen needed to breathe. Therefore, although the presence of such gases can lead to fatalities, they are not, in and of themselves, toxic.)

Throughout this section, the term "ppm" is used. It stands for "parts per million" of gas in air measured by volume. So, if the concentration of the gas is 100 ppm, then 1 m^3 of air contains 0.0001 m^3 of the undiluted gas. This is a very small volume, and can be very difficult to measure with accuracy.

Gas release modeling

The modeling of a vapor cloud following a release requires the use of highly specialized mathematical models, most of which are based on the Gaussian model, which assumes that the released gas has a normal probability distribution. Generally, the output from one of these models has a cigar shape, such as that shown in Figure 7.11, which is an elevation view for the release of the toxic gas hydrogen sulfide (H_2S).

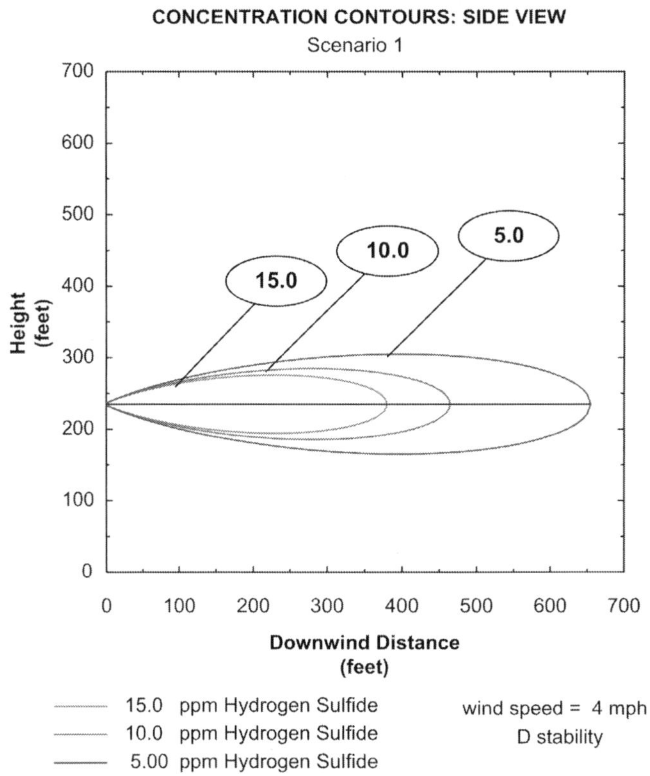

FIGURE 7.11

Gas Plume—Elevation View.

The model's output provides the following information:

- The release is at a height of 124 ft above grade.
- Three profiles are shown: one for a concentration of 15 ppm, one for 10 ppm, and one for 5 ppm.
- Using the 5 ppm profile as an example, the contour line represents the points at which the concentration of H_2S exactly equals 5 ppm. Concentrations rise toward the center of the plume, and toward the point of release.
- The furthest range of the 5 ppm plume is 650 ft.
- The plume does not show any significant change in elevation (in spite of the fact that H_2S is "heavier than air," at these low concentrations it does not rise or fall much).
- The wind speed is 4 miles/hour.
- The air has stability "D."

In general, the higher the wind speed, the more quickly the plume disperses, because the air is more turbulent. Atmospheric stability is divided into the six classes (Pasquill, 1961) shown in Table 7.5.

Table 7.5 Air Stability Classes

Class	Description
A	Very unstable
B	Unstable
C	Slightly unstable
D	Neutral
E	Slightly stable
F	Stable

Effect of toxic gases

It is difficult to predict the effect of toxic gases on the human body with any degree of certainty. Obviously, people cannot be tested directly, so data is normally taken from tests with laboratory animals. Assumptions then have to be made as to how similar a human response would be to that of the animals. Moreover, two people with very similar physiologies may react to a chemical quite differently. For example, the toxic gas H_2S has a strong odor at moderate concentrations. However, when the concentration rises above a certain point, human olfactory nerves are disabled and so the gas cannot be smelled. Thus, H_2S famously has the property that "If you can smell it you're in trouble, if you can't smell it you're in real trouble." Unfortunately, the odor threshold varies significantly from person to person, hence the sense of smell cannot be used as a reliable gauge to do with the presence of toxic gases.

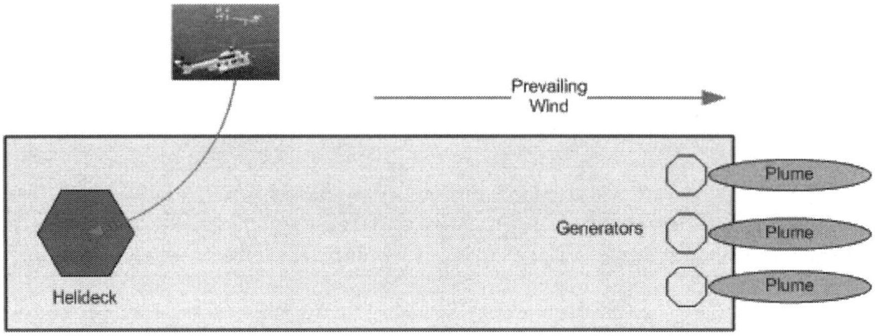

FIGURE 7.12

Location of Generators.

When information to do with the safety effects of toxic gases is not available, it is possible to use health exposure data and to extrapolate. Such an extrapolation may be very approximate, but it does provide some guidance for safety work.

Thermal plumes

Most platforms have large turbine-driven generators that use a slipstream of the produced gas. (Some platforms import power through a large umbilical cable, but that is rare.) The thermal exhaust from the gas turbines creates a plume that can pose a danger to helicopters arriving and leaving. The density of the gas in the plume is lower than that of the surrounding air. If a helicopter enters the plume, the sudden loss of lift could cause it to crash.

The best way of handling this problem is to locate the generators and helideck at opposite ends of the platform or rig, with the prevailing wind blowing the exhaust plume out to sea and away from the helideck, as shown in Figure 7.12.

In Figure 7.12, the helideck and generators are placed as far apart as possible. The prevailing wind blows the thermal plume out to sea and away from the incoming helicopter.

FIRE AND EXPLOSION ANALYSIS

Fires and explosions are obviously a major hazard on an offshore oil and gas facility. Large quantities of oil and flammable gas are present in close proximity to where personnel are working and to the LQ. Therefore, a fire and explosion study should be conducted to determine the effects of an explosion, and to make sure that adequate safeguards and controls are in place.

Fire and explosion analysis is a complex topic, and one that uses sophisticated software. Specialist support is invariably needed to carry out this work. Nevertheless, professional understanding to do with fires and explosions is still limited. For example, the initial report from the Buncefield explosion and fire that occurred in

England, December 11, 2005, states that the authors "[...] cannot explain why an ignition of gasoline vapor with no obvious source of confinement led to such a devastating explosion" (Wright, 2006).

Many of the issues regarding fires and explosions are similar to those for gas dispersion, so it is common for the two topics to be analyzed together.

Fires

Fires are often the most serious hazard faced by many process facilities. Explosions can be more devastating and will often lead to greater overall losses but they occur less frequently than fires (Kahn, 2000). Moreover, explosions are almost always followed by fires in the area that was damaged.

Fire analysis quantifies the loads on structures, equipment, or personnel arising from fires or explosions. Heat fluxes and temperatures are predicted for structures and equipment exposed to fires. This information is used to determine the response of the structure and equipment to the fire, which in turn is used to determine and evaluate risk levels and to select prevention or mitigation measures.

Flammable range

Fires require the presence of fuel and air (oxygen) along with a source of ignition. These criteria are sometimes referred to as the fire triangle. Flammable mixtures have upper and lower limits for the concentrations of the fuel in the vapor space. Below the lower flammable limit (LFL), there is insufficient flammable material for a fire to occur—the mixture is "too lean." Above the upper flammable limit (UFL), there is too much flammable material—the mixture is "too rich." The flammability limits vary according to the pressure and temperature of the mixture, and on the presence of inert components such as steam, carbon dioxide, or nitrogen.

For most flammable hydrocarbons, the LFL is around 2–5%. For gasoline vapors, the range is from a little over 1% to almost 8%. For simple alkanes such as methane and ethane, the UFL value is in the 10–15% range. Some chemicals, such as hydrogen, ethylene oxide, and acetylene, have much higher values for UFL.

Increasing the oxygen content of the flammable mixture increases the flammable range, reduces its ignition energy, and increases the energy of combustion, resulting in much more damage from the resulting explosions.

Ignition temperature/energy

Figure 7.13 can be used to illustrate ignition temperatures and flash points.

Before a flammable mixture will burn, its temperature must be at or above the flash point. If the temperature is below this point, then the vapor mixture will not burn, even if a source of ignition exists. The left line in Figure 7.13 is the flash point line.

Even if the material is above its flash point, the ignition source must be of sufficiently high temperature and must also contain sufficient energy to ignite the fuel. The minimum energy varies with type of gas and concentration; for hydrocarbon vapors it is low, for high flash point liquids, such as diesel and fuel oil, it is much higher—usually in the form of an existing fire.

If a flammable mixture is heated to a high enough temperature, it will spontaneously ignite; an ignition source such as a flame or spark is not needed. Spontaneous ignition

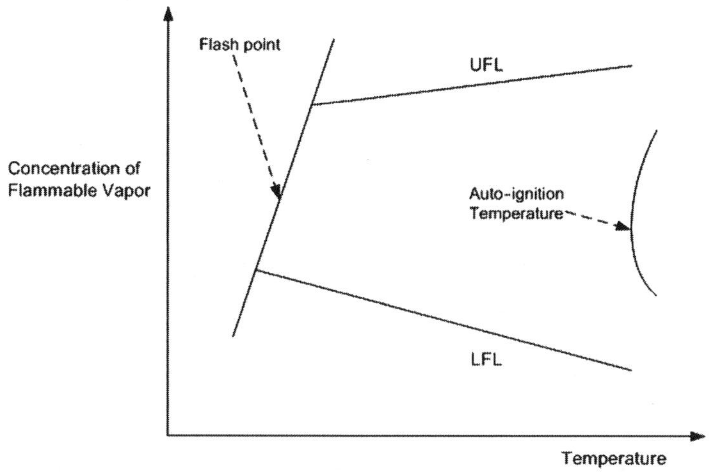

FIGURE 7.13

Flammability and Ignition Limits.

occurs at the auto-ignition temperature (AIT), which is also shown in Figure 7.13 (Zabetakis, 1965). In general, the AIT will decrease as the molecular weight of the flammable material increases.

A well-known example of auto-ignition is the operation of a diesel engine. As the diesel vapor/air mixture is compressed the temperature rises until it reaches the AIT, at which point the mixture spontaneously ignites. A gasoline engine, on the other hand, needs a spark from a spark plug to cause ignition.

Oil-soaked insulation can auto-ignite. The insulation holds the oil vapors near the heat source and prevents the oil trapped in the insulation from cooling. Furthermore, the heavy oils trapped in the insulation on very hot lines may be cracked to release lighter materials in the kerosene and diesel range, which are easier to ignite. When fed by dripping heavy oil, the oil-soaked insulation can smolder for extended periods.

Discussion on ignition issues is provided by Det Norske Veritas. Some of their conclusions are as follows:

- Natural gas released in the open will not normally auto-ignite when contacting hot surfaces at temperatures below approximately 1000 °C unless the gas remains in contact with the hot surface for a sufficiently long period of time.
- Hydrocarbon liquids will auto-ignite upon contact with hot surfaces above approximately 300 °C.
- Rotating equipment or moving parts can cause ignition due to heat generated by friction or due to static electricity discharging to earth.
- Gas from pipes that rupture due to overpressurization or corrosion is unlikely to auto-ignite because the heat generated by the rupture is unlikely to be sufficient to cause ignition. In addition, the gas has insufficient time to mix with air to form a flammable mixture at time of rupture.

- Ruptures due to impacts (e.g. dropped objects and swinging loads) can cause significant sparking, subsequently resulting in a relatively high ignition probability.
- Hot work (e.g. welding and grinding) will cause ignition of a flammable gas cloud.
- Lightning is a potential cause of both leak and ignition.
- Helicopters, boats, lifeboats, and crane engines are all sources of ignition.
- Attempts to disperse gas (e.g. by starting ventilation fans) have resulted in ignition.
- Faulty electrical equipment can cause ignition either through arcing or by the buildup of heat.

Smoke

Smoke is produced by burning hydrocarbons and solids in considerable quantities (particularly under conditions of incomplete combustion). It is also produced in large quantities by burning electrical and electronic equipment. Smoke consists of finely-divided particulate matter and suspended liquid droplets (aerosols) and can contain toxic byproducts of combustion. Smoke creates a serious hazard to personnel due to its physiological effects and reduction of visibility (most of the deaths on the Piper Alpha platform were of men in the quarters who were overcome by smoke).

During the design of a facility, smoke should be considered when selecting materials, designing ventilation systems and deciding on escape routes.

Explosions

An explosion creates an overpressure that damages or destroys buildings and equipment (most deaths and injuries are caused by flying missiles and/or building collapse rather than by the overpressure itself). Explosions often occur very quickly after initial release of gas or liquid (flashing or forming mists) and are capable of causing significant damage to a facility, sometimes where recovery is not possible. By comparison, major fire events that are not caused by an explosion do not usually cause extensive damage so quickly so there is chance of the event being controlled before structural impairment occurs.

Large explosions are capable of causing immediate critical damage, but it is also possible for relatively small explosions to result in critical escalation through damage to systems and equipment if design fails to take account of explosion hazards. The capacity of an explosion to deform structures on which equipment is supported, displace vessels and pipes through direct overpressure and drag loading, and produce serious missile hazard is what sets explosions apart from other hazards.

Some effects of explosions are as follows:

- Direct impact on systems and deformation of structure, leading to widespread loss of containment integrity as pipe and vessel connections become overstressed.
- Impairment or loss of safety systems, including ESD facilities, blowdown, and deluge systems in affected areas, and protective instrumentation. This could be from direct blast pressure, drag loads, or missile damage.

- Penetration or failure of blast barriers and fire walls, destroying protective area(s) directly or allowing subsequent fire to enter protected areas. Damage could be from blast overpressure exceeding strength of barrier, missile penetration, or failure of other structure leading to collapse of heavy equipment.
- Subsequent fire involving multiple release sites that may overwhelm fire hazard management measures, particularly if they have been damaged by the explosion (e.g. passive fire-protection coating).

Physical explosions

Physical explosions arise from a sudden release of stored energy, such as failure of pressure vessel or high-voltage electrical discharge (or even the popping of a balloon). Examples include failure of a fitting on a high-pressure gas system or failure of pressure containment in high-pressure pipes and vessels that have been physically weakened by an external event, such as a fire. The key to a physical explosion is that no chemical reactions are involved.

Vapor cloud explosions

Vapor cloud explosions are due to rapid combustion of flammable gas, mist, or small particles that generate pressure effects due to confinement; they can occur inside process equipment or pipes, buildings, and other contained areas. A vapor cloud explosion can be either a deflagration or a detonation (the distinction between deflagrations and detonations is important when deciding on whether or not to use a flame arrestor in pressure relief systems).

Deflagrations and detonations

A *deflagration* occurs when a flame front propagates by transferring heat and mass to the unburned air–vapor mixture ahead of the front. The combustion wave travels at subsonic speeds to unburned gas immediately ahead of the flame front. Flame speeds range from 1 to 350 m/s; at low speeds there is little effect from the blast overpressure, at high speeds peak overpressures can be as high as 20 times the initial pressure. Most vapor cloud explosions fall into this category.

A detonation occurs when the flame velocity reaches supersonic speeds above 600 m/s and generally in the 2000–2500 m/s range. Peak overpressures can be 20–100 times the initial pressure, with typical values of 20 bar. Detonation can be initiated either by use of a high explosive charge or from a deflagration wave that accelerates due to congestion and confinement. Certain chemicals are more prone to create detonations than normal hydrocarbons. These include ethylene, acetylene, and hydrogen.

The US EPA provides lookup tables and simple equations for some of the more common chemicals to calculate the distance of the overpressure waves. These tables are generally conservative, i.e. they predict greater impact than would be likely to actually occur. Nevertheless, they do provide a useful starting point.

Blast effects

The calculation of explosion effects is a complex topic involving many variables. Table 7.6 shows some overpressure values with typical effects.

Table 7.6 Effect of Overpressure

Overpressure (psi)	Damage
0.15–1.0	Glass failure
1.0	Person knocked down
0.4	Minor structural damage
2.0	Partial collapse of walls and roofs
3.0	Eardrum damage
3.0–4.0	Light buildings demolished; storage tanks ruptured
5.0–7.0	Complete destruction of domestic buildings; loaded railcars overturned
10.0	Total destruction of buildings
15.0	Lung damage
35.0	Fatalities

EMERGENCY SYSTEMS SURVIVABILITY ANALYSIS

The facility should be designed so that emergency systems such as the firewater deluge and lifeboats are not damaged or degraded as the result of a fire or explosion. With regard to deluge and firefighting systems, the following actions should be considered:

- Deluge valve stations should be located so that they are protected from an incident in the areas that they protect.
- Their operation may be local or remote, manual or automatic depending on the case by case assessment.
- Deluge (water spray) isolation valves should be in an accessible location under fire conditions.
- Water, particularly seawater, can have adverse effects on electrical supplies, and control of drainage water is important to avoid spreading the fire. Testing of deluge systems requires the classification of exposed electrical equipment in the protected area.
- Electrical equipment required to operate when deluge is in operation should be tested under representative conditions.
- There should be at least two firewater pumps located at different parts of the platform and with redundant drivers (usually diesel motors).

If there is a loss of containment event, the emergency shutdown system (ESD) should isolate all vessels so that the maximum leak is limited to what is in a vessel or piping system. It is particularly important that all risers are blocked in and vessels under pressure are blown down to the flare. All ventilation systems should be started, and the fire main brought up to operating pressure.

ESCAPE, EVACUATION, AND RESCUE ANALYSIS (EERA)

If all other control measures have failed, then the final step in protecting personnel is to implement the facility's emergency, escape, and rescue EER system. For this, an Escape, Evacuation and Rescue Analysis (EERA) is required.

Temporary refuge

When an emergency is declared, all workers must move in an orderly manner to their muster station, which is also a temporary refuge (TR). If the emergency escalates, the workers will be instructed to move from their muster station to the lifeboats, or other means of escape. Generally, workers will muster in the galley or similar location in the LQ. The reason for this is that many of the persons on board will be in the LQ, either eating or resting.

The TR should be provided with protection for at least one hour against fires, explosions, and smoke ingress. Therefore, the TR must be provided with an air supply, fire and gas detection, and smoke dampers. As far as is practicable, the TR should be located as far as possible from likely fires and explosions. It may be possible to place the accommodation module on mounts that will help ensure the structure's integrity if hit by a blast wave.

The TR will provide the following functions:

- It is an emergency command center (but not a control room). So, for example, it could contain a backup blowout preventer (BOP) control panel.
- Sufficient means of communication shall be provided between individuals on the installation and other installations, vessels, aircraft, and onshore.
- It provides protection for personnel and equipment. This protection can include fire and explosion walls, ventilation systems, fire-control systems, and the ability to detect fire and gas.
- It will have a sufficient supply of personal protective equipment (PPE), such as immersion suits and flotation aids.
- It will have protected access to evacuation points.
- It will have first aid capabilities.

The routes from the temporary refuge to lifeboat embarkation points should be clearly marked and well protected.

There should always be at least two muster points at different locations so that, if a person is isolated, he or she can find an alternative. For example, on a ship or barge, the primary TR is the LQ located in the superstructure at the stern of the vessel. However, if there is a fire midship and someone is in the bow area, he or she will not be able to reach the stern of the ship, so there needs to be second TR at the bow. The secondary TR should be as far away as possible from the primary muster area.

Although a TR is normally thought of as being a single, boxlike building (such as the accommodation quarters), the TR concept can be broadened to include areas where personnel can move about in safety. For example, some FPSOs have a totally-enclosed corridor running the length of the barge. This corridor provides protection from harsh weather; it also provides a safe location if someone wants to go from one end of the vessel to the other.

Toxic gas releases

If someone is working outdoors and they are exposed to a toxic gas, it is generally a good idea for that person to move indoors. As a rule of thumb, the concentration of toxic gas inside a building is around a tenth of the concentration outside. In principle, were the gas release to continue for long enough, the concentration of gas in the building would eventually increase to the level outside. However, most gas releases do not go on for a long time. After the initial puff release, the amount of emitted gas reduces, either because the inventory of stored gas is exhausted or because the affected unit is isolated by the operator or by the safety instrumentation.

Command and control

In an emergency, it is vital that a command and control system be in place. One of the many lessons learned from the Piper Alpha debacle (page 62) was that the men in the LQ, many of whom were new to the platform and had had little training, were provided with no leadership at all. They were left to their own devices, which meant that they stayed in the LQ and many succumbed to smoke inhalation.

Emergency response team

If an emergency is declared, the emergency response team (ERT) will take control of operations. There must be one person — an incident commander — who is in charge (usually the offshore installation manager, OIM). He or she is in total control over all operations and activities on the facility. Supporting the incident commander will be the ERT. The team is made up of selected employees who are specially trained to respond to fires, explosions, helicopter crashes, and other potential disasters.

The ERT will have its own muster area in which special PPE and other emergency response equipment are located.

Training

A key to the success of the system is the proper training of all persons on board. In particular, employees know where their personal muster area is located, how to reach it and what to do when they get there.

If all other control measures have failed, then the final step in protecting personnel is to implement the facility's Emergency Escape and Rescue (EER) system. For this an Escape, Evacuation and Rescue Analysis (EERA) is required.

If an emergency is declared, the following steps are followed:

- Each person moves to his or her muster station located in the Temporary Refuge (TR).
- If the situation continues to deteriorate, they evacuate from the TR, either by boat or helicopter.
- If a controlled evacuation is not possible, they are rescued, either from the sea or directly from the facility.

Emergency alarms

Audible and visual alarms are needed to inform personnel that an emergency has occurred. Where possible, the alarm should be supplemented by spoken messages over the facility's speaker system.

Escape from the incident

In an emergency, personnel must move to their designated muster point, which will be located in the TR. More than one route to helicopters and lifeboats must be present at any given time to ensure evacuation of the platform in a crisis situation. To facilitate escape from a hazardous situation, luminescent strips and heat shielding provide visibility in smoke and protection from flames. Secondary escapes such as ropes, ladders, and nets may also be available.

Escape routes

The facility design should provide escape routes from every location where a person may be present, even if those areas are not normally manned. Given that the primary escape route may be blocked, secondary or alternative escape routes should also be provided.

Escape routes should have the following features:

- The main escape route will normally be around the outside of the facility, and should be as straight and level as possible; on large platforms, there will be parallel routes at different elevations.
- The escape routes should direct personnel to the primary TR, with alternate routes to the lifeboat embarkation points.
- Escape routes should be clearly marked. Usually, this is done through the use of yellow paint on the deck, with arrows pointing in the direction of the TR. (If the deck is made of fiberglass, the colors and arrows can be "baked in.")
- The escape routes should be well illuminated at night.
- The escape routes should also be provided with plenty of signs at eye level. These signs should be designed and installed so that they can be seen when visibility is impaired (say, by smoke). They also must be illuminated so that they are visible at night (and the lights should be provided with power from an uninterruptable power supply).
- Any tripping hazards, such as steps from one section of deck to another, must be clearly marked.

T-Cards

A system for checking on the whereabouts of all personnel is needed. One widely-used system is that of T-Cards. When a person checks into the facility after arriving by boat or helicopter, he or she is assigned a T-Card. This is located at a designated spot in the TR assembly area. If an emergency is announced, each person will, on arriving at the TR, turn his or her card back to front—no one else is allowed to do this for them. Once all the cards are turned, the muster chief at that location will also call out people's names to double-check that they are present.

When all the cards are turned, and everyone is accounted for, the muster chief can then organize the evacuation of the persons for whom he or she is responsible.

Evacuation

Evacuation is the planned method of leaving the installation without directly entering the sea. Successful evacuation results in those on board the installation being transferred to an onshore location, or to a safe offshore location or vessel. This usually takes place by lifeboat or life raft. (Helicopters are used to evacuate injured personnel, but can rarely be used during a general emergency because they cannot safely approach the facility.)

The embarkation area, which is in the same general area as the muster area, should have sufficient space for all designated personnel to assemble and to board the escape craft.

Lifeboats/TEMPSC

Offshore facilities use totally enclosed motor-propelled survival craft (TEMPSC) as the primary means of water-based evacuation. Such craft possess the following features:

- They are totally enclosed, thus providing protection for the occupants against weather and sea.
- They are motorized, hence they can move away from the incident site once launched.
- They possess communications equipment, which allows them to contact rescue services.

The traditional means of launching a TEMPSC was to lower it using either fiber or wire rope (the falls) from davits. When the vessel was waterborne, the falls were released from the boat's connecting hooks. This method sometimes leads to difficulties with releasing the TEMPSC. Also, the severe wash around the base of an offshore platform can make control of the vessel very difficult, and may even lead to it being drawn under the platform. Therefore, there has been a move toward the concept of free-fall systems in which the TEMPSC falls directly into the water.

Life rafts

Offshore facilities should be provided with life rafts at different locations. These can be used if the lifeboats/TEMPSCs are not available.

Escape to the sea

If the evacuation systems are not effective, personnel may decide to escape directly to the sea.

When someone enters the water directly, he or she could die quite quickly if the water is cold. Moreover, a person may hit part of the platform structure during his or her descent.

Rescue

If personnel are obliged to leave the platform, either by TEMPSC, life raft or jumping into the water, they will have to be rescued. Usually, this is done by service boats; helicopters may not be able to come close enough, depending on the severity and type of the incident.

NON-HYDROCARBON CHEMICALS

Produced oil and gas frequently contains toxic chemicals that are hazardous to the health of those working on board, and that can cause corrosion of equipment and instrumentation. It is critical that these chemicals be handled in a proper manner.

Hydrogen sulfide

The most common toxic, non-hydrocarbon chemical found in the oil and gas industry is H_2S. This gas can cause fatalities in low concentrations. Its strong smell serves as a warning of its presence, but high concentrations of the gas can deaden the sense of smell.

The gas has a pungent odor (most texts state that H_2S smells like rotten eggs, but, with modern refrigeration, it is probably more apropos to state that rotten eggs smell like H_2S). It is easily detected by sense of smell up to values of around 100 ppm. Above 100 ppm "olfactory fatigue" can set in, and most people cannot smell the gas. Therefore, the inability to detect H_2S through the sense of smell does not prove that the gas is not present. Moreover, the ability to detect the gas by smell varies widely among individuals.

Exposures to H_2S at concentrations as low as 600 parts per million (ppm) can cause death in a matter of minutes due to paralysis of the respiratory system. Because H_2S oxidizes rapidly in the body, there are normally no permanent effects from acute exposure if the victim is rescued promptly and resuscitated before experiencing prolonged oxygen deprivation.

The ERPG-2 (Emergency Response Planning Guidelines Level 2) concentration for H_2S is 20 ppm. This is the "maximum airborne concentration below which it is believed nearly all individuals could be exposed for up to one hour without experiencing or developing irreversible or other serious health effects or symptoms that could impair an individual's ability to take protective action." In other words, a healthy worker should be able to tolerate 20 ppm of H_2S for 60 minutes without adverse effects. In practice, most electronic detectors are set for a value of 10 ppm. At this level, workers must evacuate to a safe location immediately.

H_2S can cause corrosion of stainless steels such as 316 and 410 stainless in the form of sulfide stress cracking. Copper alloys corrode rapidly in H_2S service. Upper limit values have been developed by the National Association of Corrosion Engineers (NACE MR-0175, NACE, 2008). In the gas phase, a stream is sour if the H_2S partial pressure exceeds 0.05 psia. If a single phase liquid is in equilibrium with a gas phase, where the gas phase H_2S partial pressure exceeds 0.05 psia, then that liquid is also considered to be sour.

H_2S is generally removed from the produced gas streams on the platform. If it is not removed it not only presents a hazard to workers downstream, but it can also corrode equipment. The H_2S is normally removed through use of an amine system. The removed

H_2S is then burned in the facility's generators, creating sulfur dioxide (SO_2), which is discharged to the atmosphere and can create a health hazard if the exhaust blows back on board or to another platform.

Carbon dioxide

Carbon dioxide (CO_2) is also found in many gas streams—often in quite high concentrations. Although it is not toxic, it can cause asphyxiation. Its danger is exacerbated by the fact that it has no odor. Like H_2S, carbon dioxide is frequently removed from the gas stream on the platform, with the recovered CO_2 generally being discharged to the atmosphere.

Nitrogen

Nitrogen is not, in and of itself, hazardous. After all, 79% of air is nitrogen. Nevertheless, it must be treated as a hazardous chemical. Offshore facilities frequently handle nitrogen (brought on board in liquid form) to provide inert gas for purging and maintenance operations. Like carbon dioxide, nitrogen is not toxic, but it does pose an asphyxiation hazard. Also, when in liquid form, the cold nitrogen could cause cryogenic-related problems.

If a person enters an atmosphere of nitrogen, he or she can lose consciousness without any warning symptoms in as little as 20 seconds. Death can follow in 3–4 minutes. One deep breath of 100% nitrogen will be fatal because breathing is stimulated and controlled by the concentration of carbon dioxide in the lungs. 100% nitrogen displaces the carbon dioxide completely, so breathing stops.

Personnel should not work in or enter atmospheres containing less than 19.5% oxygen, unless equipped with a self-contained breathing apparatus or breathing air mask. This is also true of rescue personnel who can be overcome by the same oxygen-deficient atmosphere as the initial victim.

Treatment chemicals

In addition to produced chemicals, offshore facilities generally use a range of chemicals for treatment of the oil, gas, and produced water. In particular, hydrate formation in subsea pipelines is controlled with chemicals such as methanol and glycol. A commonly used specialty chemical is low-dosage hydrate inhibitor (LDHI).

These treatment chemicals are toxic if ingested in large quantities, and they are usually flammable. But they are not nearly as hazardous as some of the chemicals handled in a typical refinery or chemical plant. Nevertheless, an analysis of the chemicals that are on board should be carried out. Issues to consider include the following:

- How the chemicals are stored (totes or tanks, usually).
- The potential for a chemical leak and its consequences (fire or the creation of toxic fumes).
- The potential for spilled liquid accumulating (e.g. on one vessel a large spill of methanol would flow back toward the LQ, creating the potential for a fire in that area).

- How chemical fumes are controlled, e.g. with a seawater spray deluge.
- Whether people are properly trained to handle a spill.
- What would happen were the supply of that chemical to be interrupted.

GREENHOUSE GAS EMISSIONS

Offshore platforms generate significant quantities of greenhouse gases—particularly carbon dioxide from the generators, which are often powered by produced gas. The FSA may include a requirement that these emissions are identified and measured. In addition, some jurisdictions require that companies report on their greenhouse gas emissions. One example is the Australian *Offshore Petroleum & Greenhouse Gas Storage (Environment) Regulations*, published in 2009.

NOISE AND VIBRATION

Noise is airborne sound energy within a broad range of frequencies that has the potential to cause either discomfort or injury to people working in the area. Specifically, noise needs to be controlled in order to:

- Protect the hearing of the personnel who work at the facility
- Ensure that warning and alarm signals can be heard
- Allow for normal speech between workers (direct, radio and telephone)
- Provide a quiet location for workers to relax and sleep

Onshore facilities also need to consider the impact of noise on the local community.

Noise is generally measured in decibels weighted to the "A" scale which approximates the ear's response to noise. Hence, most readings are given as dB(A) or dBA. This single number covers a wide range of frequencies. A decibel is a logarithmic value to the base 10. Therefore, if the decibel value is increased by ten, the noise exposure is doubled. More sophisticated noise analyses will consider the actual pressure on a person's eardrum and will consider the frequency of the noise waves.

The level of noise drops off quickly the further away a person is from the noise source. Therefore, when making decisions about the control of noise, it is important to determine where the workers are likely to be located. As a starting point, a distance of 1 m is often assumed.

Discussion about noise will generally also cover the topic of vibration. Not only can vibration affect the health of workers in the same way as noise, does, it can also cause equipment damage and failure.

Because space is almost always severely limited offshore, workers and accommodation areas are generally close to equipment and other noise sources. Hence, noise management and control often plays an important role in the design of a facility—particularly with regard to layout.

Sources of noise

There are many sources of noise, but the following are usually of the greatest concern:

- *High-pressure letdown valves*. If a control valve is used to reduce the pressure of a gas stream, considerable noise can be generated.
- *Flares*. The noise from the flame at the flare tip can be considerable.
- *Rotating equipment*. Pumps and compressors can generate considerable levels of noise.
- *Air intakes and vents*. Fans, vents, burners, and air-conditioning can all generate high levels of noise.
- *Electrical equipment*. Transformers and motors can be noisy.
- *Helicopters*.

Noise problems offshore are often compounded by the fact that the noise waves can be reflected off walls and other surfaces, and may actually be amplified.

Allowable noise and vibration levels

A common rule of thumb is that a person must not be exposed to noise levels in excess of 85 dB(A) over an 8-hour period, or 88 dB(A) over a 12-hour period (HSE, 2001). Both these numbers are for a time-weighted average (TWA).

Some increase in the allowable noise level may be permitted for special operations. For example, sound levels of 115 dB(A) may be permitted for short periods of time during start-up, shutdown, regeneration, and maintenance. However, workers must wear hearing protection during such operations.

The upper limit for allowable sound under any circumstances, including emergencies, is usually set at around 115 dB(A).

If the equipment is onshore and the requirements of the local community need to be considered, then noise limits may be different for different times of day or night and for workdays or weekends, particularly if the facility is located in a built-up neighborhood.

In all cases, it is necessary to define the work areas that are covered. One company states that a work area is any position in which a person may be present and which is no less than 1 meter from equipment surfaces, or any position where a worker's ear may be exposed to noise in the normal course of his duty. It includes platforms, walkways, and ladders.

Table 7.7 provides representative guidance as to allowable noise levels in different situations.

The Occupational Safety and Health Administration (OSHA) uses the "5 dBA rate" for managing duration. This standard states that noise exposure duration needs to be cut in half if the noise level increases by 5 dB(A).

Where equipment emits an intermittent or fluctuating noise (say, when it is being depressured or blown down), noise limits such as those shown below can be applied:

- 8 hours: 85 dB(A)
- 4 hours: 88 dB(A)

Table 7.7 Allowable Noise Levels dB(A)

Area Description	Maximum Level
Areas in workshops and machinery buildings where communication is required Workshops for light maintenance	70
Kitchens	60
Workshop offices Control rooms, not continuously manned Computer rooms	60
Continuously manned control rooms Open plan offices Social rooms, changing rooms, wash places, and toilets	50
Offices and conference rooms	45
Radio and communication rooms	45
Accommodation	40

- 2 hours: 91 dB(A)
- 1 hours: 94 dB(A)

Where the equipment emits noise fluctuating in a more complicated manner, the equivalent continuous sound level can be calculated according to the method given in EEMUA 140 (EEMUA, 1985).

Noise control

The desirable hierarchy of noise-control systems is shown in Figure 7.14.

Engineering controls

Engineering controls are the best way of managing noise because they are inherently safe—if the noise is not created in the first place then there is no need to put barriers and controls in place.

Pathway control

If it is not possible to engineer out the noise problem, the next step is to restrict the flow of noise from where it is created to where the people are located. The restriction can be achieved either through the use of absorbents or by deflecting surfaces or the combination of both.

Noisy machinery is sometimes placed inside a small containment building or acoustic enclosure. Acoustic enclosures may be either of the "close-fitting" or "walk-in" (large) type. The use of acoustic enclosures is generally not a preferred solution to noise problems because they create a confined space and have the potential for a buildup of a flammable or explosive cloud should the equipment leak. Also, the containment structure is likely to make maintenance and inspection more difficult

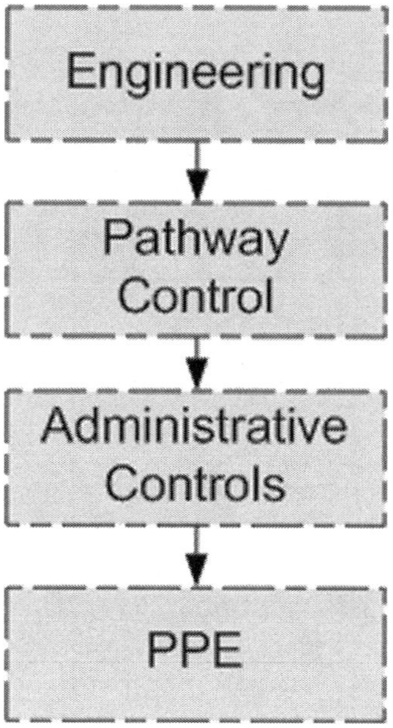

FIGURE 7.14

Noise-Control Hierarchy.

(a confined space entry will be required), and it will increase the size of the equipment's footprint.

Noise transmission is generally controlled either through the use of barriers or sound-absorbing materials (or a combination of both), as illustrated in Figure 7.15.

The source noise hits a wall that is retaining a layer of absorbent material. The noise is attenuated as it passes through the absorbent. It then hits a barrier wall. Some of the noise is reflected back through the absorbent material, and some is transmitted to where the workers are located.

Other pathway controls include the following:

- Control valves can have an internal multipath geometry that lowers noise levels. It is important to ensure that this feature does not increase the chance of the valve being plugged by solid materials.
- Silencers can be installed on engines and motors. It is important that any internal parts that may inadvertently break loose do not partly or completely block the silencer outlet or damage downstream equipment (e.g. compressors).
- In-line silencers can reduce the noise associated with fluid flow. Silencers can either contain baffles and chambers, or porous material such as stainless steel wool. Guidance to do with the design of silencers is provided by the Engineering and Equipment Users Association (EEMUA, 1988).
- Insulation or lagging can be put on equipment to reduce noise levels, typically by 5–20 dB(A).

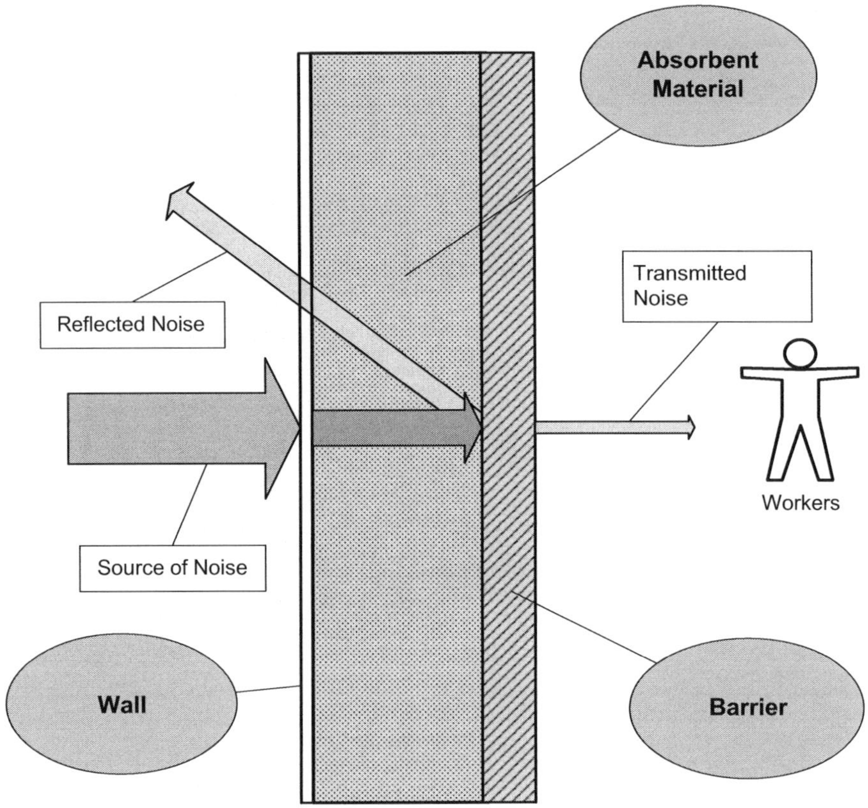

FIGURE 7.15

Barriers and Sound Absorption.

Vibration control can be achieved in the following ways:

- Repair or modify the equipment that is causing the vibration.
- Modify the equipment's mounting.
- Put the equipment on steel or air springs.
- Install isolation hangars for pipe that is connected to vibrating equipment.

Administrative controls

Administrative controls aim to keep people at a safe distance from the noise source. Examples include signs and notices, barricades, procedures, and training.

PPE

The last resort is to require employees to wear PPE such as ear plugs and ear muffs. (Many facilities require that anyone working outside wear ear plugs as a precautionary principle, regardless of their location or what they are doing.)

HUMAN FACTORS ENGINEERING

Human factors is defined as the science of making plant, equipment, and processes robust enough to deal with human error without adverse effects. The term "ergonomics" is used in Europe, but is generally considered to apply more narrowly to the man—machine interfaces and interactions. When applied to the design of offshore facilities, the term "human factors engineering" (HFE) is sometimes used in order to identify the topic as being to do with engineering design rather than human behavior. Whatever term is used, the topic covers the interface between humans and technology, including plant equipment, instrumentation, software, and work processes. The aim is to optimize human efficiency, effectiveness, health, safety, and well-being, within the context of overall system performance operability, reliability, and maintainability.

In spite of its obvious importance, HFE tends to be something of an orphan child during the design of facilities. The topic rarely has a well-defined position within safety management systems in the way that, say, hazards analysis does. This lack of clarity is illustrated by the following quotation from the general section of API RP 75.

> *Human factors may be considered in the design and implementation of the company's Safety and Environmental Management Program.*

The above statement provides no guidance as to the scope of a human factors analysis, when it is to be done, or how it links to the other elements of SEMP.

RELIABILITY, AVAILABILITY, AND MAINTAINABILITY

RAM analyses are not, strictly speaking, part of an FSA. However, they are often carried out at the same time as the FSA for the following reasons:

- The results of a RAM analysis could affect important design decisions such as the number of spare equipment items or the types of equipment selected. These decisions will, in turn, affect the overall safety of the facility.
- There are strong similarities between the techniques used for RAM work and for PRA. For example, techniques such as FTA calculate the likelihood of an "event" taking place. The event could be a fatal accident, or it could be the failure of a compressor. The same calculation method can be used for both cases.
- Both safety and reliability quantitative analyses make use of the same equipment reliability and maintainability data.

In spite of the similarities between the techniques used for RAM analysis and for safety assessments, their goals are fundamentally different. When discussing RAM, it is possible to talk about optimum reliability, i.e. the point at which a dollar spent on improving reliability leads to less than a cost-averaged dollar in benefits, as illustrated in Figure 7.16. With safety, however, there is no real optimum value: all incidents are unacceptable. As explained with the discussion of ALARP (page 22), no company or government agency is likely to commit to an acceptable level in the number of deaths or serious injuries.

Figure 7.16 illustrates the concept of optimum reliability. It shows that, as funds are initially expended on improved reliability, the incremental revenue is greater than the money spent (when factored over the normal capital investment period). However,

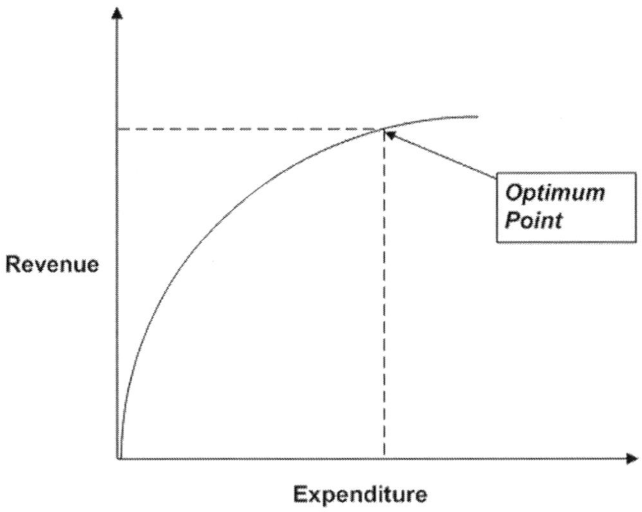

FIGURE 7.16

Reliability Payout.

there is an optimum point, above which a dollar spent on improved reliability generates less than a dollar in life cycle incremental revenue (i.e. the slope of the curve becomes less than unity).

In practice, there is rarely sufficient data to be able to develop a curve such as Figure 7.16 with precision. Nevertheless, it is useful to keep in mind that a reliability program is not, in and of itself, its own justification. It has to demonstrate that an investment in reliability will lead to an increase in profits.

Human reliability

The topic of human reliability, i.e. getting people to do the right things at the right time, presents management with a dilemma. Reliability should and can be quantified, and the management should "follow the numbers"—both engineering numbers and dollar numbers. However, this is very difficult to do with human beings because they are not machines, and cannot be analyzed and modeled in the same manner as pumps, tanks, and instruments. Most analysis of human performance falls in the realms of psychology and sociology rather than engineering.

Yet human error and human reliability are a profoundly important part of system reliability—indeed it could be argued that, in the limit, all reliability problems are to do with human error. If a piece of equipment fails to operate properly, then possible human causes include the following:

- The item was not properly specified by the design engineer.
- The maintenance supervisor did not organize an effective inspection program.
- A designer may have calculated loads and stresses incorrectly.
- Management failed to implement a preventive maintenance program.
- The operator may have failed to follow instructions.

Human reliability analysis (HRA) is used to determine the probability that a task or activity will be completed successfully within a required time period, and that no other human action that could be detrimental to system performance will take place. An HRA analysis can also help identify areas where potential improvements can be made.

Errors can either be of commission or omission. Errors of commission typically involve failure to follow procedures, taking a short cut or making an (incorrect) assumption about the validity of an instrument reading. Errors of omission often occur during the response phase of an incident. For example, an operator may fail to isolate a tank that has already started to overflow.

A potentially serious human error occurs when an operator or supervisor does not realize that he or she has exceeded a safe operating limit. Not realizing how far out of control the operations have become, he or she decides to fight the problem rather than shut down and bring the facility operations to a safe state.

Human reliability analysis

An HRA usually consists of the following stages:

- Identification of key tasks
- Task analysis of key tasks
- Human error identification
- Human error analysis
- Human reliability quantification—often using technique for human error rate prediction (THERP), as discussed below.

THERP

One method for analyzing human reliability is a straightforward extension of PRA—in the same way that equipment can fail, so can a human make mistakes and slips. One technique for predicting human error rates is the THERP, which was developed in the 1950s. As with other PRA techniques, THERP models can use either point.

A THERP analysis considers different types of error, such as not following an instruction, choosing a wrong switch or skipping a step in a sequence of activities, and forecasts the error rate for each of these tasks. If a person can make more than one type of error when carrying out a task, then the probabilities are added to one another. For example, when opening a valve an operator may

- Open the wrong valve;
- Skip the step altogether; or
- Open it only partway.

If the respective probabilities for these errors are 0.01, 0.03, and 0.03, then the overall error rate is 0.07 (excluding second-order terms). It is also possible to factor in recovery rates. For example, if the wrong valve is selected, then there may be a 40% probability that the operator will recognize and correct the error while there is still time, thus reducing the overall probability of error to 0.6×0.01, or 0.006.

Valve criticality analysis

A useful human factors exercise is to conduct a valve criticality analysis. Typically, manually operated valves are organized into categories such as "operated daily," "operated occasionally," and "used only during turnarounds." The category selected will determine the location and accessibility of the valve—with those that are used daily being provided with easy access that does not require any stretching or the use of ladders or other temporary equipment.

Reliability data

The Offshore Reliability Data Bank (OREDA) provides data on the reliability of offshore components and systems. It was created by eight oil and gas companies. They collect and exchange reliability data with one another. Although OREDA is based on North Sea data, it is widely used in offshore location around the world. It also covers some onshore equipment.

REFERENCES

Civil Aviation Authority (UK), 2008. CAP 437 Offshore Helicopter Landing Areas - Guidance on Standards.

EEMUA. Engineering and Equipment Users Association 140, 1985. Noise Procedure Specification.

EEMUA. Engineering and Equipment Users Association 161, 1988.

HSE (United Kingdom Health & Safety Executive). Noise and Vibration. Offshore Technology Report. 2001/068.

HSE (United Kingdom Health & Safety Executive), 2002. Offshore Hydrocarbon Releases Statistics.

HSE (United Kingdom Health & Safety Executive), 2003. PARLOC 2001: The Update of Loss of Containment Data for offshore Pipelines. 5th edn. 2003.

International Maritime Organization (IMO), April 2002. Guidelines for Formal Safety Assessment (FSA) for Use in the IMO Rule-making Process.

ISO 31010, 2009. Risk Management - Risk Assessment Techniques.

ISO 17776:2000(E), 2000. Petroleum and Natural Gas Industries — Offshore Production Installations - Guidelines on Tools and Techniques for Hazards Risk Assessment.

Juran, J.M. 1951. Quality Handbook. Republished McGraw-Hill 1999.

Khan, F., et al. October 2000. Avoid the domino effect via proper risk assessment. Chemical Engineering Progress.

NRC (Nuclear Regulatory Commission), 1975. Reactor Safety Study: An Assessment of Accident Risks in U.S. Commercial Nuclear Power Plants.

Pasquill, F. 1961. The Estimation of the Dispersion of Windborne Material. Meteorological Magazine vol. 90 (No. 1063).

SINTEF. Offshore Reliability Data Handbook 5th Edition, Volume 1 - Topside Equipment, Volume 2 - Subsea Equipment. 2009.

United Kingdom Offshore Operators Association (UKOOA), 2003. Fire and Explosion Guidance.

Vesely, W. 1981. Fault Tree Handbook.

Wright, T. September 2006. Still searching for Buncefield answers. Hydrocarbon Processing.

Zabetakis, M. 1965. Flammability characteristics of combustible gases and vapors. Bulletin 627, Bureau of Mines.

CHAPTER

Offshore safety developments

8

CHAPTER CONTENTS

INTRODUCTION

The previous seven chapters of this book have demonstrated how much the offshore safety management business has changed and matured in the quarter century since the Piper Alpha accident. Yet, as the Deepwater Horizon and Montara events illustrate, more progress needs to be made, particularly with respect to process and technical safety. The industry has not yet "arrived." No doubt ongoing application of the Safety and Environmental Management Systems (SEMS) and safety case techniques will lead to further improvement, yet there may also be a need to make a step change, and to find new ways of improving safety.

Some thoughts as to how the industry is changing and the effect that these changes may have on system safety are provided in this chapter. Topics discussed are as follows:

- Leadership and management
- Follow the rules
- Need for technical expertise
- Convergence of standards
- Spill response
- Long-tail or Black Swan events

LEADERSHIP AND MANAGEMENT

The word *management* occurs throughout this book around 943 times, and the same word is to be found in the key phrases: Safety and Environmental *Management* Program

Ian Sutton: Offshore Safety Management. DOI: 10.1016/B978-1-4377-3524-6.00008-3
Copyright © 2012 Ian Sutton. Published by Elsevier Inc. All rights reserved.

and Safety and Environmental *Management* Systems. Yet it is useful to draw a distinction between management and leadership, and to examine how improved leadership could have prevented, or at least made less serious, many of the accidents discussed in Chapter 2.

Looked at this way, a manager is someone who develops and implements systems for activities such as operating procedures, training and equipment integrity. In principle, a management system will, to a certain extent, run itself, regardless of the specific individuals involved. Leadership, on the other hand, is highly dependent on the personality of the person in charge. Leaders need to make difficult decisions, often under considerable time pressure—all the while realizing that they are the officers in charge of troops who may at any time be facing the equivalent of enemy fire. *Leaders must lead*.

Leadership is often described in terms of creating systems, and then encouraging all employees and contract workers to adhere to those systems. For example, a recent article states, "Leaders must take the responsibility to influence behaviors regarding process safety" (Louvar, 2011). While this statement is certainly true, such a function is more to do with management than leadership. A leader recognizes that they are the officer in charge of troops whose lives may be threatened, and they must be willing to take very tough decisions under stressful and dangerous conditions.

Arnold (2010) states, "We do not need another level of analysis and documentation to improve safety—we do need to create organizations which encourage their people to take risks for safety." In other words, leaders must be willing to use their judgment to take the right actions in real time, even if those actions lead to a large economic loss.

In the case of the Piper Alpha catastrophe, it is thought that seven men died in the initial explosion. Yet the final death toll was 167. The fire that followed the initial explosion should have burnt itself out within 30 minutes—*had the incoming riser valves been closed*. But those valves were not closed and so the fire burned for hours. Consequently, an additional 160 men died of burns, smoke inhalation, and drowning.

The death toll was so high due to lack of leadership. Piper was a hub platform that received oil and condensate from other platforms in the area, one of which was Tartan. When Piper exploded, the managers on Tartan could see perfectly well what was happening—Piper was in line of sight, and visibility (even though it was nighttime) was good. Yet Tartan continued to pump condensate into the raging inferno on Piper because, in the words of the OIM, "no one told me to stop." He did not take the initiative to shut down production, maybe because he was afraid of getting "into trouble" if he had done so.

Neither was there adequate leadership on Piper itself. Many of the men on board were in quarters at the time of accident (because it was nighttime). They had not been properly trained, so they stayed together until most of them succumbed to smoke asphyxiation. Had there been stronger leadership on the platform, an officer in charge would have led them to a location where they would have had a better chance of escaping.

It would appear as if lack of leadership was also a factor in the Deepwater Horizon incident (although the final reports were not available at the time of writing). The men on the rig were experiencing serious problems for two days, and some of the decisions that they made—such as the use of a long string—seem to have been made as a result of economic pressures. Yet no one at senior level (either on the rig or the onshore office) had the courage to call a time-out (knowing that doing so would cost $1 million per day). They were not willing to "take a risk for safety."

The Blackbeard event (Chapter 2) showed how leadership should work. The members of the drilling crew were worried about the high temperatures and pressures that they were observing. They decided to call on senior management for guidance. In the end, the decision went to the top of the company—and the decision was to declare a dry hole (which, of course, it was not). They were willing to abandon a $100 million plus investment in order to avoid a calamitous accident.

FOLLOW THE RULES

Many of the incidents discussed in Chapter 2 could have been avoided had those involved simply followed the rules. In the case of the Santa Barbara blowout, the persons in charge did not obey the standards set by the regulators. The same can be said about the recent Montara blowout. The report to do with that event states,

> *Well control practices approved by the delegate of the Designated Authority, the Northern Territory Department of Resources, most likely would have been sufficient to prevent the Blowout if PTTEPAA had adhered to them and to its own Well Construction Standards.*
>
> *These difficulties were compounded by the fact that senior PTTEPAA personnel had only limited experience of batch drilling and batch tieback operations and did not fully comprehend the implications of such operations.*

While it is easy to say that people must follow the rules, once more leadership enters into the mix. People do not skirt the rules for the sake of it—they do so usually because they are trying to make more production more quickly or, as apparently was the case with PTTEPAA, management did not know the rules thoroughly enough. A strong leader will resist the pressures to cut corners or to make ill-informed decisions. He or she will insist that the rules, procedures, and standards be followed unless it can be demonstrated that doing so would give rise to an unsafe situation.

LACK OF TECHNICAL EXPERTISE

One engineering company used the slogan shown below as a bumper sticker.

There's No Substitute For
Knowing What You're Doing

Anyone who has led a process hazards analysis (PHA) will immediately understand the significance of the above phrase. The literature to do with PHAs is replete with lengthy discussions as to the techniques to be used, the manner in which the discussion is to be organized, and the guidewords to be followed. However, if the PHA team consists of people who really understand the facility—people who *do* know they they are doing—then they will quickly identify the important issues and come up with sensible findings, regardless of the formal organization of the study.

A phrase sometimes used in the petroleum industry to describe the loss of experienced personnel is "Big Crew Change" (Gell, 2008). The phrase is mainly used with regard to the large number of baby-boomer generation engineers and technical experts who will be retiring after the year 2010. These experienced personnel are not being replaced by sufficient younger people with comparable technical skills and experience. For example, it has been reported (Corrigan, 2005) that:

> *Between 1983 and 2002, the number of US petroleum engineers declined from 33,000 to 18,000 ....*

There is always company knowledge that cannot be brought in off the street. If a person has worked at a company for many years, particularly in a specialist department, and he or she leaves without having trained a replacement, there is almost certain to be a loss of "corporate memory." People brought in from the outside may be very knowledgeable, but they cannot possibly know all of the history and background as to why things operate the way they do at this particular location.

A particular concern is that the number of experienced people in many organizations has been reduced, and the gaps between the layers of supervision and management have increased. Hence, many process issues that were previously handled quickly and effectively on a semiformal basis by experienced personnel who knew each other very well, and who also had an intimate knowledge of the processes for which they were responsible, now must be handled in a more formal manner.

In other areas of plant operations, the loss of personnel is compensated for by using increasingly sophisticated computer systems. For example, a Distributed Control System (DCS) can carry out many of the functions previously performed by several operators. Similarly, sophisticated design software lets one engineer carry out calculations that previously had to be done by a team. Yet there remain certain actions that have to be carried out by people; the loss of skilled personnel in these situations represents a true loss. No computerized system can develop the intuitive insights of an experienced operator or design engineer.

There is no easy solution to the dilemma of loss of experience. Once an experienced person has left the organization, it will take time to build up qualified replacements. Training courses and seminars can help—but they are likely to be too generic and high level to provide the knowledge and experience that is needed.

All that can be said is that companies should be cautious about losing their experienced people by "offering them the package" and they should allow plenty of time for the junior people to absorb the knowledge of their more senior counterparts.

CONVERGENCE OF STANDARDS

Much ink has been spilled (or toner cartridges consumed) in discussions regarding the relative merits of Safety and Environmental Management Program (SEMP)/SEMS and safety case approaches to improving offshore safety. To some extent, the choice of approach will partly depend on local rules and regulations, and also on environmental conditions. In the North Sea, for example, much of the equipment is enclosed to protect it from the weather. This is not the case for Gulf of Mexico facilities, where weather conditions are generally mild (except when a hurricane comes through). The presence of enclosed spaces on North Sea platforms increases the likelihood of a confined vapor cloud explosion, and thus affects the types of safety analyses that are carried out. Also, going overboard in the North Sea is a very serious matter, whereas the waters of the Gulf of Mexico are far more benign, as demonstrated in the Mariner incident in which 13 men entered the water, but no one was injured.

In spite of these differences, the reality is that there is a good deal of overlap between the safety case and SEMP/SEMS approaches. Both result in the development of a Safety Management System, and both are basically goal-based and nonprescriptive, although SEMP and SEMS do reference a large number of prescriptive standards, mostly from the American Petroleum Institute (API).

Not only are there strong similarities between the two systems, but they also appear to be converging. In the Gulf of Mexico, the original approach was to manage the safety of large numbers of relatively simple, shallow water platforms through the use of generic standards. This approach made sense when the platforms are similar to one another and when a single platform is not all that expensive. It does not make sense to prepare a safety case for each and every one of these facilities. However, with the move to deep water drilling and production, the situation has changed. Now each drill rig or platform is a unique design, and its cost is enormous—usually well over a billion dollars. This means that it makes sense to develop a facility-specific safety analysis, in other words to prepare a document much like a safety case.

In the North Sea, the trends are in the other direction. The production of oil and gas is declining quite rapidly. Consequently, many of the major oil companies have sold their producing platforms to smaller organizations, some of which are little more than holding companies. These new owners do not have the financial or organizational resources to develop and manage safety cases. Instead, they say, "Just tell us what to do." They favor a more prescriptive approach to managing safety, just as companies did in the early days of work in the Gulf of Mexico.

SPILL RESPONSE

If the worst happens, and there is a big release of oil or gas to the ocean, industry needs to be able to respond quickly and effectively. The Deepwater Horizon incident demonstrated that such a response capability was not in place—it took industry three months to cap the leak from the damaged wellhead and to direct the flow of oil to a safe location. The National Commission report to the President says the following about that event:

> *Just as the events of April 20, 2010 exposed a regulatory regime that had not kept up with the industry it was responsible for overseeing, the events that unfolded in the subsequent weeks and months made it dismayingly clear that neither BP nor the federal government was prepared to deal with a spill of the magnitude and complexity of the Deepwater Horizon disaster.*

Industry in the Gulf of Mexico responded by developing two projects: Helix and the Marine Well Containment System (MWCS). Helix is based on the equipment and systems that were used following Deepwater Horizon. MWCS is a brand new system.

(The National Commission report to the President notes that neither of these projects will necessarily address the next crisis. For example, if a drill rig were to fall directly on top of the wellhead, these systems would not be able to access the spill site. Moreover, these systems are limited to a water depth of 10,000 ft, even though industry is developing the technology to drill in much deeper water.)

Helix Fast Response System

The Helix Energy Solutions Group, headquartered in Houston, Texas, USA, is an international offshore energy company that was instrumental in developing the containment systems used for the Macondo well blowout. Based on that experience, the company developed the Helix Fast Response System (HFRS). Participants in the project pay a retainer fee and a day rate to HFRS, who will provide the equipment needed to control a leak.

The system uses two existing vessels: The Helix Producer I and Q4000, a semi-submersible rig. A stack can, consisting of a well cap and intervention riser system, will transfer oil from the damaged blowout preventer (BOP), wellhead or tree to a riser connected to Q4000. From there the oil will be pumped through a flexible riser to Helix Producer I that can carry out some processing. Oil would then be transferred to a tanker that would take it to an onshore facility.

Marine Well Containment System

The MWCS is being developed by a not-for-profit organization that is sponsored by many of the large oil companies, including ExxonMobil (who are taking the lead), Royal Dutch Shell, Conoco Phillips, and Chevron.

The system uses two ships that are in normal tanker service in the Gulf of Mexico. The ships are designed such that, in the event of an emergency, they can go to port, have the capture equipment installed on them, and then move to the location of the leaking oil well. Oil will be stored on the ships—which have been converted to FPSOs—and transferred to shuttle tankers, as needed.

The subsea part of the system will be connected with the damaged well and will have the capability of totally capturing the leak, even if the wellhead is damaged. An initial subsea response system was put into service in February 2011.

LONG-TAIL EVENTS

> *When anyone asks me how I can best describe my experience of nearly 40 years at sea, I merely say uneventful. Of course there have been winter gales and storms and fog and the like, but in all my experience, I have never been in an accident of any sort worth speaking about. I have seen but one vessel in distress in all my years at sea. I never saw a wreck and have never been wrecked, nor was I ever in any predicament that threatened to end in disaster of any sort.*
>
> *E.J. Smith, Captain of the* Titanic

At times it seemed as if industry had a response to the Deepwater Horizon catastrophe on same lines as that of Captain Smith. Many people were stunned that the event could have occurred, and managers from other companies were quick to point out that it could not have happened in their organization. Yet it did happen, and it became an industry-wide issue.

The nuclear power industry seems to suffer from the same "it cannot happen here" mind-set. Probabilistic risk models, such as those discussed in Chapter 7, demonstrate that the likelihood of a nuclear power plant catastrophe is miniscule. Yet it seems as if such events occur every decade or so—each time caused by factors that no one had planned for.

- The Three Mile Island accident (1979) was attributed largely to failures in human factors and the ability of management and supervision to understand what was going on.
- The consequences of the Chernobyl catastrophe (1988) were greatly exacerbated by the decision not to construct a containment building.
- The Fukushima Dai-ichi event (2011) is still unfolding. But already it is clear that a central factor in its cause was the decision not to build a sea wall high enough to handle a once-in-a-thousand year tsunami.

Each of the above incidents had causes that will probably never repeat. Yet the fact remains that a serious nuclear incident has occurred roughly once every 10–15 years. It would be naïve to believe that some other unique, hard-to-predict nuclear power event is not going to take place within the next two or three decades. Similarly, leaders in the oil industry must recognize that catastrophic events are going to occur more frequently "than they should."

Black Swans

The best-selling book *The Black Swan* by Nassim Nicholas Taleb showed how catastrophic events seem to occur with surprisingly high frequency and how they were always a surprise to the people affected. Although his work was carried out in the financial world, his lessons apply to any type of industry that has the potential for catastrophic events, which are sometimes referred to as "long-tail" events, which means that a larger share of events rests in the tail of a probability distribution than for a normal Gaussian distribution.

Taleb defines a Black Swan as having the following three features:

(1) The event is a surprise.

(2) The event has a major impact.

(3) After the fact, the event is rationalized by hindsight, as if it *could* have been expected.

Deepwater Horizon fits all of the above criteria. There is no question that, at the time, the event was a surprise. It certainly had a major impact. And, by now, the event has been thoroughly analyzed and explained.

Short-term thinking

The natural response would be for companies to make the necessary investments to protect themselves against long-tail events, but, in the words of Taleb:

> *Unfortunately, most managers are rewarded on short term gains, i.e. on "flat earth" programs, not on preparing for a rare, event that cannot be defined or explained and that may not happen for a very long time.*

Sutton (2010) provides a fictional illustration of the above dilemma.

> Three years ago, under pressure from head office, we downsized our engineering department from 20 to 15 professionals. The savings helped the plant meet its financial goals (and also improved my bonus). Head office was so pleased with the result that a year later we cut another five engineering professionals. Some of the nagging reliability problems we have had since then can probably be attributed to these cuts, but we did improve profitability once again, and I have to accept that our operation is hanging in there.
>
> Given our success in cutting costs, senior management decided to relocate 4 of the remaining 10 professionals to head office (200 km away) so as to reduce travel costs and office expenses. So now we are down to 6 on-site engineers from the 20 we had just 3 years ago.
>
> I'm worried that we may be heading for a major accident because we have cut our technical support so drastically. Maybe I should raise a warning flag. However, I can't be sure, so I'll just have to trust to luck. After all, serious accidents happen only rarely, and I expect to be assigned to a new position within a year or two. Then these worries become someone else's problem.

Short-term thinking encourages just-in-time management and the minimization of redundancy. Just as a bank needs a certain percentage of its money to be held in reserve, so the offshore industry needs to have reserves of skilled personnel and redundant equipment to protect against the "unknown unknown." (The development of the spill recovery systems discussed in the previous section is an example of this insurance mentality.)

Much of the short-term thinking approach can be attributed to cost pressures, and the need to "do more with less." In his book *Lessons from Longford* (Hopkins, 2000) the author states, "when we extend the causal network [for an accident] far enough, market forces and cost cutting pressures are almost invariably implicated."

Referring to the Titanic disaster, Brander (1995) writes:

> *Most of the discussion of the accident revolves around specific problems. There was the lack of sufficient lifeboats (enough for at most 1200 on a ship carrying 2200). There was the steaming ahead at full-speed despite various warnings about the ice-field. There was the lack of binoculars for the lookout. There were the poor procedures with the new invention, the wireless (not all warnings sent to the ship reached the bridge, and a nearby ship, the operator abed, missed Titanic's SOS) [...]*

What gets far less comment is that most of the problems all came from a larger, systemic problem: the owners and operators of steamships had for five decades taken larger and larger risks to save money—risks to which they had methodically blinded themselves. The Titanic disaster suddenly ripped away the blindfolds and changed dozens of attitudes, practices, and standards almost literally overnight.

CONCLUSIONS

This book has attempted to provide an overview of the topic of offshore safety management—with a particular focus on technical and process safety, and the avoidance of catastrophic events. The development of process safety management systems such as SEMS and technical safety systems, such as the Formal Safety Assessment part of safety cases, has been discussed in depth. Although it is difficult to pin down how much improvement such systems have generated, there can be little doubt that they have been effective at reducing the number and severity of offshore accidents.

Yet more progress needs to be made, and it is likely that new ways of thinking will be needed to supplement the work that is already being done with SEMS and safety cases. It is suggested in this chapter that two quite disparate approaches are required.

The first is to encourage leaders within organizations to actually lead, as distinct from managing. They need to understand that they must, at times, "take a risk for safety." The second approach is to do with following the rules and minimizing the number of Black Swans. And it is in this area that regulations play an important role.

Regulations encourage a "Follow the Rules" mentality because they give government agencies the power to enforce standards and recommended practices. Managers will respond to the fear that they may be cited in an audit, or, even worse, face severe personal consequences if there is a serious accident.

In addition, if a regulation is written so as to force companies to plan for rare events, then everyone is working on the same terms. This is analogous to environmental regulations. Such regulations are expensive to comply with, and key people seem to spend too much time "filling out forms." But without the regulations some companies would choose not to bother about pollution and so would have an unfair advantage over their more conscientious competitors.

Therefore, it is suggested here that it is not whether regulations such as SEMS and safety cases are desirable—but whether they provide a sensible set of rules that genuinely help improve process and technical safety. If they can help prevent future Piper Alphas and Deepwater Horizons, then they are justified.

REFERENCES

Louvar, J.F. June 2011. How to Prevent Process Accidents. Process Safety Progress.

Sutton, Ian. 2010. Behavior-based Safety. www.stb07.com/occupational-safety/behavior-based-safety.html.

Hopkins, A. 2000. Lessons from Longford: The Esso Gas Plant Explosion. CCH Australia Ltd.

Taleb, N.N. 2007. The Black Swan. Random House, New York, USA.

Arnold, Ken. September 2010. Production Operations - The need for a Safety Case. SPE Seminar Houston.

Brander, Roy. 1995. The Titanic Disaster: An Enduring Example of Money Management vs. Risk Management.

Corrigan, M. November 2005. Innovative, Flexible Strategies Needed to Overcome Human Capital Deficit. Journal of Petroleum Technology.

Gell, C., Schilling, L. 2008. Empowering Young Professionals to Lead After "The Big Crew Change". Society of Petroleum Engineers 115341.

References

API (American Petroleum Institute) RP 75, 2004. Development of a Safety and Environmental Management Program for Offshore Operations and Facilities.

BOEMRE, 2010. 30 CFR Part 250. Oil and Gas and Sulphur Operations in the Outer Continental Shelf—Safety and Environmental Management Systems.

Broadribb, M. April 2008. 3 years on from Texas City. 4th Global Congress on Process Safety.

Bromwich, M. May 2011. Future of Offshore Oil and Gas Development in the U.S. Offshore Technology Conference.

Civil Aviation Authority (UK), 2008. CAP 437 Offshore Helicopter Landing Areas - Guidance on Standards.

CSB (U.S. Chemical Safety and Hazard Investigation Board): Regulatory Approaches to Offshore Oil and Gas Safety, *Public Hearing*, December 2010.

Cullen, D. 1990. The Public Inquiry into the Piper Alpha Disaster. Department of Energy, HMSO Cm 1310, London.

DOCEP (Department of Consumer and Employment Protection (Western Australia)), September 2005. Mine Safe vol. 14 (No. 3).

EEMUA. Engineering and Equipment Users Association 140, 1985. Noise Procedure Specification.

EEMUA. Engineering and Equipment Users Association 161, 1988.

Fewtrell, L., Bartram, J. (eds.) Water Quality Guidelines, Standards and Health. London: IWA World Health Organization.

Gell, C., Schilling, L. 2008. Empowering Young Professionals to Lead After "The Big Crew Change". Society of Petroleum Engineers 115341.

Glenn, David G: Job Safety Analysis. Its Role Today, *Professional safety*, March 2011.

Heiler, K. October 2005. Is the Australian mining industry ready for a safety case regime? 31st International Conference of Safety in Mines Research Institutes.

Hopkins, A. 2000. Lessons from Longford: The Esso Gas Plant Explosion. CCH Australia Ltd.

HSE (United Kingdom Health & Safety Executive). Noise and Vibration. Offshore Technology Report. 2001/068.

HSE (United Kingdom Health & Safety Executive), 2002. Offshore Hydrocarbon Releases Statistics.

HSE (United Kingdom Health & Safety Executive), 2003. PARLOC 2001: The Update of Loss of Containment Data for Offshore Pipelines. 5th edn. 2003.

International Maritime Organization (IMO), April 2002. Guidelines for Formal Safety Assessment (FSA) for Use in the IMO Rule-making Process.

ISO 31010, 2009. Risk Management - Risk Assessment Techniques.

ISO 17776:2000(E), 2000. Petroleum and Natural Gas Industries — Offshore Production Installations - Guidelines on Tools and Techniques for Hazards Risk Assessment.

Juran, J.M. 1951. Quality Handbook. Republished McGraw-Hill 1999.

Khan, F., et al. October 2000. Avoid the domino effect via proper risk assessment. Chemical Engineering Progress.

Kinsella, Kevin. September 2010. UK Offshore Safety Cases: Lessons Learnt. SPE Seminar, Houston.

Lees, F. 2004. Loss Prevention in the Process Industries. Elsevier, Burlington.

Louvar, J.F. June 2011. How to Prevent Process Accidents. Process Safety Progress.

Maguire, R. 2006. Safety Cases and Safety Reports. Ashgate.

MOD (Ministry of Defence (UK)), December 1996. Safety Management Requirements for Defence Systems. Defence Standard 00:56, Issue 2.

MOD (Ministry of Defence (UK)), December 2004. Safety Management Requirements for Defence Systems Part 1. Interim Defence Standard 00:56, Issue 3.

Montara Commission of Inquiry, June 2010. Commonwealth of Australia.

National Commission on the BP Deepwater Horizon Oil Spill and Offshore Drilling. January 2011.

NRC (Nuclear Regulatory Commission), 1975. Reactor Safety Study: An Assessment of Accident Risks in U.S. Commercial Nuclear Power Plants.

NSC (2009). Accident Prevention Manual for Industrial Operations. fifth ed. 2009.

Parker W: *SEMS Workshop Opening Statement*, 2009 0902, Offshore Operators Committee.

Pasquill, F. 1961. The Estimation of the Dispersion of Windborne Material. Meteorological Magazine vol. 90 (No. 1063).

Pitblado, R. October 2008. Global process industry initiatives to reduce major accident hazards. 11th Annual Symposium. Mary Kay O'Connor Process Safety Center.

Rasche, Tilman. October 2001. Development of a Safety Case Methodology for the Minerals Industry - A Discussion Paper. University of Queensland.

Sutton, Ian. 2010a. Behavior-based Safety. www.stb07.com/occupational-safety/behavior-based-safety.html.

Sutton, Ian S. 2010b. Process Risk and Reliability Management. Elsevier, Oxford, England.

Taleb, N.N. 2007. The Black Swan. Random House, New York, USA.

UK HSE, 1997. Offshore Installations (Prevention of Fire and Explosion, and Emergency Response) Regulations, 1995 (PFEER). Approved Code of Practice and Guidance, L65. HSE Books.

United Kingdom Offshore Operators Association (UKOOA), 2003. Fire and Explosion Guidance.

United States Steel Corporation: *Principles of Accident Prevention in United States Steel*, third ed., Pittsburgh, Pennsylvania, 1968, United States Steel Corporation.

Vesely, W. 1981. Fault Tree Handbook.

Wright, T. September 2006. Still searching for Buncefield answers. Hydrocarbon Processing.

Zabetakis, M. 1965. Flammability characteristics of combustible gases and vapors. Bulletin 627, Bureau of Mines.

Index

Page references followed by "f" indicate figure, and by "t" indicate table.

O

P

R